Springer Series in Statistics

Perspectives in Statistics

Advisors
P. Bickel, P. Diggle, S. Fienberg, K. Krickeberg,
I. Olkin, N. Wermuth, S. Zeger

Springer

New York
Berlin
Heidelberg
Barcelona
Hong Kong
London
Milan
Paris
Singapore
Tokyo

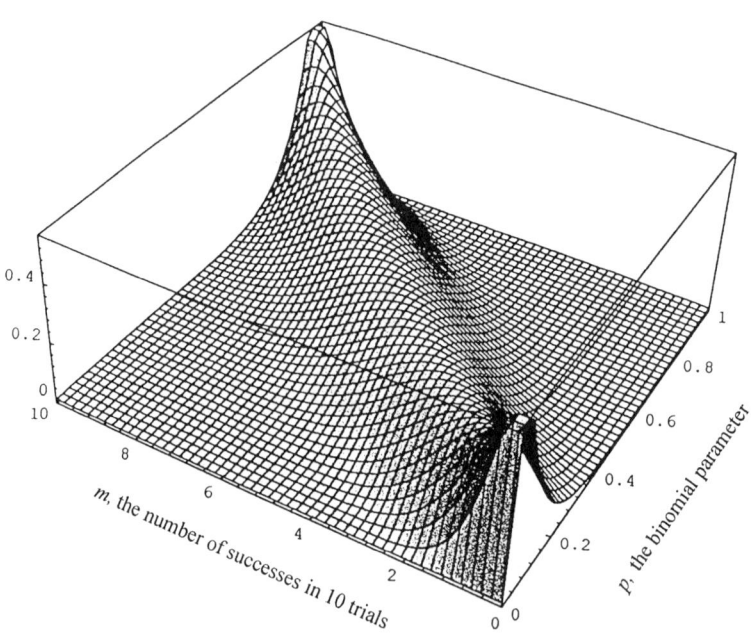

Frontispiece: A 3-D *Mathematica* plot of the probability/likelihood diagram for a binomial parameter in Thiele (1889). For every value of the binomial parameter p the binomial distribution is plotted, the values at non-integral values of m, the number of successes, being found by interpolation (see our discussion of Thiele's article). In the orthogonal direction the curves are therefore the likelihood functions for p for each m. Note that this diagram, unlike Thiele's, does not extend into the regions $m < 0$ and $m > 10$.

H.A. David
A.W.F. Edwards

Annotated Readings in the History of Statistics

 Springer

H.A. David
Statistical Laboratory and
    Department of Statistics
Iowa State University
102 Snedecor Hall
Ames, IA 50011-1210
USA
hadavid@iastate.edu

A.W.F. Edwards
Gonville and Caius College
Cambridge CB2 1TA
UK
awfe@cam.ac.uk

With 3 illustrations.

Library of Congress Cataloging-in-Publication Data
David, H.A. (Herbert Aron), 1925–
    Annotated readings in the history of statistics / H.A. David, A.W.F. Edwards.
      p.   cm. — (Springer series in statistics)
    Includes bibliographical references and index.
    ISBN 0-387-98844-0 (hardcover: alk. paper)
      1. Mathematical statistics—History.   I. Edwards, A.W.F. (Anthony William Fairbank),
    1935– II. Title. III. Series.

QA276.15 .D39   2001
519.5′09—dc21
                                                       00-041977

Printed on acid-free paper.

Production managed by Michael Koy; manufacturing supervised by Erica Bresler.
Typeset by Archetype Publishing, Inc., Monticello, IL.
Printed and bound by Edwards Brothers, Inc., Ann Arbor, MI.
Printed in the United States of America.

9 8 7 6 5 4 3 2 1

ISBN 0-387-98844-0   SPIN 10725505

Springer-Verlag New York Berlin Heidelberg
A member of BertelsmannSpringer Science+Business Media GmbH

Dedicated to
Anders Hald
Professor Emeritus of Statistics
University of Copenhagen, Denmark
Author of
A *History of Probability and Statistics and Their Applications Before* 1750
and
A *History of Mathematical Statistics from* 1750 *to* 1930

Contents

Preface   ix

Sources and Acknowledgments   xiii

General Note on the Translations   xv

The Introduction of the Concept of Expectation:
Comments on Pascal (1654) 1

The First Formal Test of Significance:
Comments on Arbuthnott (1710) 7

Coincidences and the Method of Inclusion and Exclusion:
Comments on Montmort (1713), N. Bernoulli (1713),
and de Moivre (1718) 19

On the Game of Thirteen by P. R. de Montmort 25

Letter from Nicholas Bernoulli to Montmort on the Game of Thirteen,
reproduced in Montmort (1713, p. 301) 31

The Doctrine of Chances,
Problem XXV by A. de Moivre 32

The Determination of the Accuracy of Observations:
Comments on Gauss (1816) 37

The Determination of the Accuracy of Observations by C. F. Gauss 41

The Introduction of Asymptotic Relative Efficiency:
Comments on Laplace (1818) 51

On the Probability of Results Deduced by Methods of any Kind
from a Large Number of Observations by P. S. Laplace 57

The Logistic Growth Curve: Comments on Verhulst (1845) 65

Mathematical Investigations on the Law of Population Growth
by P.-F. Verhulst 69

Goodness-of-Fit Statistics: The Distributions in Normal Samples of (a) the
Sum of Squares About the Population Mean, (b) the Circular Sum
of Squares of Successive Differences, and (c) the Circular Serial
Correlation Coefficient: Comments on Abbe (1863) 77

On the Conformity-to-a-Law of the Distribution of Errors in a Series
of Observations by E. Abbe 81

The Distribution of the Sample Variance Under Normality:
Comments on Helmert (1876b) 103

The Calculation of the Probable Error from the Squares of the
Adjusted Direct Observations of Equal Precision and Fechner's
Formula by F. R. Helmert 109

The Random Walk and Its Fractal Limiting Form:
Comments on Venn (1888) 115

Estimating a Binomial Parameter Using the Likelihood Function:
Comments on Thiele (1889) 129

Yule's Paradox ("Simpson's Paradox"): Comments on Yule (1903) 137

Beginnings of Extreme-Value Theory:
Comments on Bortkiewicz (1922a) and von Mises (1923) 145

Range and Standard Deviation by L. von Bortkiewicz 151

On the Range of a Series of Observations by R. von Mises 155

The Evaluation of Tournament Outcomes:
Comments on Zermelo (1929) 161

The Evaluation of Tournament Results as a Maximization Problem
in Probability Theory by E. Zermelo 167

The Origin of Confidence Limits: Comments on Fisher (1930) 187

Appendix A. English Translations of Papers and Book Extracts of
Historical Interest 203

Appendix B. First (?) Occurrence of Common Terms in
Statistics and Probability 209

Subject Index   247

Name Index   251

Preface

Interest in the history of statistics has grown substantially in recent years and the subject is now covered by a number of excellent books. S.M. Stigler's *The History of Statistics* (1986) gives an overview up to 1900 while Anders Hald's two encyclopedic volumes *A History of Probability and Statistics before 1750* and *A History of Mathematical Statistics from 1750 to 1930*, published in 1990 and 1998, provide detailed mathematical discussion of the major contributions up to 1930. Hald's books have removed Isaac Todhunter's *A History of Probability* from the pedestal which it occupied for a century and a quarter and rendered Karl Pearson's *Lecture Notes* of mainly historical interest themselves. Journal papers have appeared on specific topics, especially in the series "Studies in the History of Probability and Statistics" in *Biometrika* and in the long sequence of papers in *Archive for the History of the Exact Sciences* by O. Sheynin. The two volumes of reprinted papers, mostly from *Biometrika*, issued in 1970 and 1977 have proved particularly valuable. More recently, many important papers published since 1900 have been reprinted with commentaries in the three-volume *Breakthroughs in Statistics* (1992–1997). Stigler's *Statistics on the Table* (1999) provides illuminating vignettes.

In addition, specialized books have appeared on particular topics, such as A.I. Dale's *A History of Inverse Probability* (1991, 1999) and R.W. Farebrother's *Fitting Linear Relationships* (1998). The pioneering book on the early period, F.N. David's *Games, Gods and Gambling* (1962), has in some particulars been superseded by works such as Oystein Ore's *Cardano, the Gambling Scholar* (1980), Ian Hacking's *The Emergence of Probability* (1975), and A.W.F. Edwards's *Pascal's Arithmetical Triangle* (1987). Papers on the history of probability and statistics are now commonplace in the regular journals, modern history being particularly well represented in *Statistical Science*.

Such a rapid expansion of a field has its own dangers, however. Gaps are left which need to be filled and interpretations become accepted which need to be reviewed. Our purpose in the present work is to make easily accessible to the English-speaking reader a number of contributions, spanning three centuries, which have hitherto been relatively neglected (often because they have not previously been translated) or which seem to us to merit reassessment. The commentaries on the contributions vary considerably in style and length according to the purpose of the selection. Thus, the brief extracts from Pascal's writings in 1654 attract a long commentary because we believe they deserve more attention than they have hitherto received. By contrast, we have included Abbe's 1863 paper more in order to make it readily available, than to comment on it at length, since we are able to refer the reader to the admirable commentary by M.G. Kendall. The articles themselves also vary greatly in length, the two longest, by Abbe and Zermelo, being remarkable papers by authors much better known for their contributions outside statistics.

Much of the important work in statistics was, of course, written in languages other than English. We feel that there is a real dearth of translations. A special feature of this book therefore arises naturally: the large number of translated articles. The translations are largely the work of the authors and include writings by Pascal, Montmort, Gauss, Laplace, Verhulst, Abbe, Helmert, Thiele, Bortkiewicz, Mises, and Zermelo.

Whereas the reasons for our selections will mostly be evident from a reading of the associated commentaries, our reasons for not including certain other pieces may not be so clear, but will primarily be because they are already available as modern reprints, with commentary. Obvious examples are Thomas Bayes's famous paper from 1764, Karl Pearson's paper on goodness-of-fit (1900), and "Student's" on the t-test (1908). In an appendix we have listed for the convenience of the reader references to translations of a number of important historical pieces. In a second appendix entitled *First (?) Occurrence of Common Terms in Statistics and Probability* the first author has integrated, corrected, and supplemented his previous work on the subject. For the authors of most of our contributions biographical information is readily available in *Leading Personalities in Statistical Science* (1997), but for the remainder we have added some biographical details. Good likenesses for most authors are available on the History of Mathematics archive created by O'Connor and Robertson, at *http://www-history.mcs.st-and.ac.uk/history*.

Some of the articles featured in this volume were used by the first author in a short course on the history of statistics, recently given at Iowa State University. The course emphasized reading of the original literature.

We are indebted to Anders Hald, not only for his excellent books, but also for specific comments on some of our selections. Comments were also gratefully received from John Aldrich, M.S. Bartlett, and G.A. Barnard. Special thanks are due to Steffen Lauritzen for his translation from the Danish of the Thiele extract and for allowing us to use it in advance of publication of the whole book in translation. It is a pleasure also to acknowledge helpful responses, to a variety of queries, from Oscar Sheynin, Sarah Nusser, and Ruth David. This is apart from help, acknowledged in Appendix B, from the many who contributed to First (?) Occurrences of Statistical Terms. John Kimmel, editor of the Springer Series in Statistics, provided crucial early encouragement and was most helpful throughout.

We thank the Iowa State University Department of Statistics for general support. Specifically, Jeanette LaGrange provided secretarial assistance and Sharon Shepard expertly prepared the LaTeX typescript.

The second author gratefully acknowledges a Royal Society of London Research Grant which enabled him to visit Iowa.

H.A. David, Ames, Iowa
A.W.F. Edwards, Cambridge, UK
March 2000

Sources and Acknowledgments

Pascal, B. (1654). Extracts from a letter to Fermat and from the *Traité du triangle arithmétique*. Reprinted in J. Mesnard (ed.), (1970). *Oeuvres complètes de Blaise Pascal, deuxième partie, volume I: Oeuvres diverses*. Desclée De Brouwer, Bruges. Translated by A.W.F. Edwards.

Arbuthnott, J. (1710). An argument for Divine Providence, taken from the constant regularity observ'd in the births of both sexes. *Phil. Trans.*, **27**, 186–190. Reproduced by kind permission of the Master and Fellows of Trinity College, Cambridge.

Montmort, P.R. de (1713). *Essay d'analyse sur les jeux de hazard*, 2nd edn. *Revûe et augmentée de plusieurs lettres*, pp. 130–138. Facsimile reprint. Chelsea, New York (1980). Translated by H.A. David.

Bernoulli, N. (1713). Letter (in Latin) to Montmort, p. 301 of Montmort (1713). Extract translated by H.A. David.

Moivre, A. de (1718). *The Doctrine of Chances*. Pearson, London, pp. 59–63. Reproduced by kind permission of the Syndics of Cambridge University Library.

Gauss, C.F. (1816). Bestimmung der Genauigkeit der Beobachtungen. *Z. Astron u. Verwandte Wiss.*, **1**, 185–196. *Werke*, **4**, 109–117, Königliche Gesellschaft der Wissenschaften, Göttingen (1880). Translated by H.A. David.

Laplace, P.S. (1818). *Théorie analytique des probabilités, deuxième supplément*, Courier, Paris. *Oeuvres*, **7**, 613–623, Imprimerie Royale, Paris (1847) Translated by H.A. David.

Verhulst, P-F. (1845). La loi d'accroissement de la population. *Nouveaux mémoires de l'académie royale des sciences et belles-lettres de Bruxelles*, **18**, 1–38. Translated by H.A. David.

Abbe, E. (1863). Ueber die Gesetzmässigkeit in der Vertheilung der Fehler der Beobachtungsreihen. Habilitationsschrift, Jena. Reprinted in *Gesammelte Abhandlungen*, **2**, 55–81 (1904). Translated by H.A. David.

Helmert, F.R. (1876). Die Genauigkeit der Formel von Peters zur Berechnung des wahrscheinlichen Fehlers directer Beobachtungen gleicher Genauigkeit. *Astron. Nachr.*, **88**, columns 113–132. Columns 120–124 translated by H.A. David.

Venn, J. (1888). *The Logic of Chance*, 3rd edn. Macmillan, London, pp. 107–118.

Thiele, T.N. (1889). Forelæsninger over almindelig Iagttagelseslære: Sandsynlighedsregning og mindste Kvadraters Methode. Reitzel, København. Translated by Steffen L. Lauritzen.

Yule, G.U. (1903). Notes on the theory of association of attributes in statistics. *Biometrika*, **2**, 121–134. Pages 132–134 reproduced by kind permission of the Council of the Royal Statistical Society, London, and the Master, Fellows, and Scholars of St. John's College, Cambridge.

Bortkiewicz, L. von (1922). Variationsbreite und mittlerer Fehler. *Sitzungsberichte der Berliner Math. Gesellschaft*, **21**, 3–11. Pages 3–6 translated by H.A. David with the kind permission of B.G. Teubner, Stuttgart, Publishers.

Mises, R. von (1923). Über die Variationsbreite einer Beobachtungsreihe. *Sitzungsberichte der Berliner Math. Gesellschaft*, **22**, 3–8. Translated by H.A. David with the kind permission of B.G. Teubner, Stuttgart, Publishers.

Zermelo, E. (1929). Die Berechnung der Turnier-Ergebnisse als ein Maximumproblem der Wahrscheinlichkeitsrechnung. *Math. Zeit.*, **29**, 436–460. Translated by H.A. David.

Fisher, R.A. (1930). Inverse probability. *Proc. Camb. Phil. Soc.*, **26**, 528–535.

General Note on the Translations

As far as possible we have adopted the policy of literal translation, but with such modifications as seem necessary to help the modern reader. Also, in an effort to preserve the flavor of the original, we have retained its style of equations and its notation except when altogether too confusing nowadays. An example is the longtime use of μ to denote standard deviation.

Any equations occurring in our commentaries are labeled (a), (b), ... so as to distinguish them from numbered equations in the original. Labels (A), (B), ... have been added to equations not numbered in the originals but referred to in the commentaries.

The Introduction
of the Concept of Expectation

Comments on Pascal (1654)

The notion of the expected value of a gamble or of an insurance is as old as those activities themselves, so that in seeking the origin of "expectation" as it is nowadays understood it is important to be clear about what is being sought. Straightforward enumeration of the fundamental probability set suffices to establish the expected value of a throw at dice, for example, and Renaissance gamblers were familiar enough with the notion of a fair game, in which the expectation of each player is the same so that each should stake the same amount. But when more complicated gambles were considered, as in the Problem of Points, no one was quite sure how to compute the expectation. As is often the case in the development of a new concept, the interest lies not so much in the first stirrings which hindsight reveals but in the way its potential came to be realized and its power exploited. In the case of expectation, Pascal clarified the basic notion and used it to solve the Problem of Points.

This problem, also known as the "division problem," involves determining how the total stake should be divided in the event of a game being terminated prematurely. Suppose two players X and Y stake equal money on being the first to win n points in a game in which the winner of each point is decided by the toss of a fair coin. If such a game is interrupted when X still lacks x points and Y lacks y, how should the total stake be divided between them? In the middle of the sixteenth century Tartaglia famously concluded "the resolution of such a question is judicial rather than mathematical, so that in whatever way the division is made there will be cause for litigation." A century later the correct "mathematical" solution was derived by three different arguments during a correspondence between Blaise Pascal and Pierre de Fermat in the summer of 1654 (for a full account see Edwards, 1982). One of the methods, advanced by Pascal in his letter of July 29 part of which we reproduce here in translation, involves computing expectations recursively. As Hacking (1975) remarked in his chapter "Expectation," "Not until the correspondence between Fermat and Pascal do we find expectation well understood."

The key advance was Pascal's understanding that the value of a gamble is equal to its mathematical expectation computed as the average of the values of each of two equally probable outcomes, and that this precise definition of value lends itself to recursive computation because the value of a gamble that one is certain to win is undoubtedly the total stake itself. Thus if the probabilities of winning a or b units are each one half, the expectation is $\frac{1}{2}(a+b)$ units, which is then the value of the gamble. Expected value, that is, the probability of a win multiplied by its value, is here revealed for the first time as an exact mathematical concept that can be manipulated.

The word itself ("*expectatio*" in Latin) is due to Huygens (1657) in his *De ratiociniis in ludo aleæ*, and it is usually asserted that the concept is his too, but the balance of the available evidence is well summarized by Hacking (1975) who wrote "He [Huygens] must have heard the gist of the solutions [Pascal's and Fermat's] but he went home and worked them out for himself." For Huygens had spent July to September 1655 in Paris, associating with the circle of scholars who later formed the *Académie Royale des Sciences*, during which, although he did not meet Pascal himself on that occasion, he had the opportunity to hear about Pascal's work on probability problems from Roberval, with whom Pascal had discussed his and Fermat's solutions in detail. In *De ratiociniis* Huygens's first Proposition is "If I may expect either a or b and either could equally easily be obtained, then my expectation should be said to be worth $\frac{1}{2}(a+b)$," exactly as Pascal had argued. In Propositions II and III this is generalized, and in Propositions IV to IX Huygens treats the Problem of Points for two and then three players in just the same way that Pascal had.

In his Preface to *De ratiociniis* Huygens modestly says "It should be known that already some of the most outstanding mathematicians in France have discussed this calculus, lest anyone should attribute to me the unde-served glory for its first invention." But he adds that these mathematicians, when challenging each other with difficult problems, did not reveal their methods, and that therefore he has had to develop the subject from first principles (*à primis elementis*). One such principle is surely the use of ex-pectation in the Pascalian manner, and it is difficult to imagine how a conversation between Roberval and Huygens about Pascal's work could have avoided mentioning it, or indeed what else they could have discussed about the work. In any event, through correspondence with Carcavi in the summer of 1656 Huygens learned that his method was the same as Pascal's, as a result of which he added Proposition IX, so that before *De ratiociniis* went to press he was aware of Pascal's calculus of expectations (Edwards, 1987, Ch. 8, fn. 8).

For an extended account of the emergence of expectation which lays greater emphasis on the contemporary importance of aleatory contracts and other legal doctrines, see Daston (1988). However, although this back-ground might have affected Huygens's exposition, it does not negate the evidence for Pascal's influence, and in her account Daston (like F.N. David, 1962, before her) overlooks the information about Pascal's thinking pro-vided by his *Traité du triangle arithmétique*. This work had much less prominence than Huygens's because it was not circulated until 1665 and did not receive the attention of James Bernoulli in *Ars Conjectandi*, but it was written and printed in 1654 and thus supplies valuable information about Pascal's thinking then.

For a description of Pascal's *Traité* see Edwards (1987); for an account of the role of expectation in "Pascal's wager" *infini-rien* see Hacking (1975); for a detailed analysis of "*expectatio*" in Huygens's writing see Freudenthal (1980); and for a remarkable forgotten solution to the Problem of Points from about 1400, in which the value of each play is computed on the principle that at each toss the gain to a player if he wins must equal the loss if he loses, see Schneider (1988).

We here print in translation (1) an extract from Pascal's letter of July 29, 1654 to Fermat and (2) an extract from the section of the *Traité du triangle arithmétique* entitled *Usage du triangle arithmétique pour déterminer les partis qu'on doit faire entre deux joueurs qui jouent en plusieurs parties*. In the letter he develops the recursive argument applied to expected values in order to find the correct division of the stake-money and thus computes the "value" of each successive throw. In the *Traité* the same idea is more formally expressed, and in particular, in the extract we give, Pascal gives as a principle the expectation when the chances are equal. He does not, however, use any word equivalent to "expectation," although in the preceding paragraph he does use "*l'attente du hasard*" to describe the position which the players are agreeing to replace by each receiving his expectation instead of continuing to play.

Note on the translation and sources. The letter of which the following is an extract was preceded by others (which have not survived) and therefore presumes familiarity with the Problem of Points. Pascal refers to the settling of each point as a "game" ("*partie*"), but since the English word "game" usually means competing for several points (as in tennis, for example) we have translated "*partie*" as "throw." It is implicit in his letter that each game or throw has two equally probable outcomes, as with the toss of a fair coin. Pascal's letter to Fermat was first published in 1679 while the *Traité* was published in Paris in 1665 (Desprez). For our translation we have used the edition of Pascal's works by Mesnard (1970).

References

Daston, L. (1988). *Classical Probability in the Enlightenment*. Princeton University Press.

David, F.N. (1962). *Games, Gods and Gambling*. Griffin, London.

Edwards, A.W.F. (1982). Pascal and the Problem of Points. *International Statistical Review*, **50**, 259–266. Reprinted in Edwards (1987).

Edwards, A.W.F. (1987). *Pascal's Arithmetical Triangle*. Griffin, London, and Oxford University Press, New York.

Freudenthal, H. (1980). Huygens' foundations of probability. *Historia Mathematica*, **7**, 113–117.

Hacking, I. (1975). *The Emergence of Probability*. Cambridge University Press.

Huygens, C. (1657). *De ratiociniis in ludo aleæ*. In F. van Schooten *Exercitationum mathematicarum libri quinque*, pp. 517–534. Elsevier, Leyden.

Mesnard, J. (ed.) (1970). *Oeuvres complètes de Blaise Pascal, deuxième partie, volume I: Oeuvres diverses*. Desclée De Brouwer, Bruges.

Schneider, I. (1988). The market place and games of chance in the fifteenth and sixteenth centuries. In C. Hay (ed.), *Mathematics from Manuscript to Print*, 1300–1600, pp. 220–235. Clarendon Press, Oxford.

Pascal to Fermat, 29 July 1654 (part).

Sir,

Impatience afflicts me as well as it does you and, although I am still confined to bed, I cannot resist telling you that yesterday evening I received through M. de Carcavy your letter on the division of the stakes, which I admire more than I can say. I do not have the leisure to reply at length but, in a word, you have found the two divisions, for the dice and for the throws [*problem of points*] with perfect justice; I am entirely satisfied, for I do not doubt any more that I was correct, seeing the admirable accord in which I find myself with you.

I admire the method for the throws much more than that for the dice. I have seen many people find that for the dice, such as M. le Chevalier de Méré, who it was who proposed the question to me, and also M. de Roberval; but M. de Méré has never been able to find the just value in the case of the throws, nor any route for getting it, so that I have found myself the only one who has known this result.

Your method is very sound and is the one which first occurred to me in this research, but because the tedium of the combinations is excessive I have discovered a shortcut, or, rather, another method which is much shorter and clearer, which I hope to be able to tell you about now in a few words. For I should like from now on to open my heart to you, if I may, such is the pleasure of realizing that we are in agreement. I see that the truth is indeed the same in Toulouse as in Paris.

Here, more or less, is how I go about finding the value of each of the throws when two players play, for example, for three throws [*to be the first to win three throws*] and each has staked 32 pistoles:

Suppose that the first player has won two and the other one; they now make another throw whose chances are such that if the first wins he gains all the money which is at stake, that is to say 64 pistoles; but if the other wins the score is two all, so that if they want to separate it follows that each should recover his stake, namely 32 pistoles.

Consider then, sir, that if the first wins, 64 belong to him; if he loses, 32. Thus if they want not to chance another throw and to separate without making it, the first should say, "I am certain to get 32 pistoles even if I lose; but as for the other 32, perhaps I get them and perhaps you do — the chances are equal. Let us therefore share these 32 pistoles equally between us and in addition allot me the 32 which are mine for sure." He will then have 48 pistoles and the other player 16.

Let us now suppose that the first player has won two throws and the other none, and that they throw again. The chances of this throw are such that if the first player wins it he takes all the money, 64 pistoles, but if the other wins they are back at the preceding case in which the first has two wins and the other one.

Now we have already shown that in this case 48 pistoles are due to the person who has two wins, so that if they do not want to make this throw he ought to argue thus, "If I win it, I shall gain everything, that is, 64 pistoles, but if I lose it, I will still be entitled to 48. Therefore give me the 48 to which I am entitled even if I lose and let us share equally the remaining 16 because there is the same chance that you will win them as I will." Thus he will have 48 and 8, that is 56 pistoles.

Finally let us suppose that the first player has won only one throw and the other none. You see, sir, that if they make a new throw the chances are such that, if the first wins, he will have two wins to none and, by the preceding case, 56 are due to him; but if he loses, they are one all in which case 32 pistoles are due to him. Then he should argue, "If you do not wish to play, give me the 32 pistoles of which I am certain and let us share the remainder of the 56 equally. 56 take away 32 leaves 24; divide 24 in half, take 12 and allot me 12 which with 32 will make 44."

Now in this manner you see that, by simple subtraction, for the first throw 12 pistoles of the other's money belong to the winner; for the second another 12; and for the last, 8.

Pascal, *Traité du triangle arithmétique,* **1654 (extract).**

The first principle leading to a knowledge of the way in which one should make the division is as follows:

If one of the players finds himself in the position that, whatever happens, a certain sum is due to him whether he loses or wins and chance cannot take it from him, he should not divide it but take it all as is his right, because the division has to be proportional to the chances and as there is no risk of losing he should get it all back without division.

The second principle is this: if two players find themselves in the position that if one wins he will get a certain sum and if he loses then the sum will belong to the other; and if the game is one of pure chance with as many chances for the one as for the other and thus no reason for one to win rather than the other, and they want to separate without playing and take what they are legitimately due, the division is that they split the sum at stake into half and each takes his half.

First corollary

If two players play a game of pure chance on condition that, if the first wins, a certain sum is returned to him, and if he loses, a lesser sum is returned to him; and if they wish to separate without further play and each take what belongs to him, the division is that the first takes what he is due if he loses plus half the difference between what comes to him if he wins and what comes to him if he loses.

For example, if two players play on condition that, if the first wins he carries off 8 pistoles, and if he loses he carries off 2, I say that the division is that he takes the 2 plus half of the excess of 8 over 2, that is to say, plus 3, since 8 exceeds 2 by 6, half of which is 3.

Second corollary

If two players play under the same condition as before, I say that the division can be made in the following way which comes to the same thing: that one adds the two sums of the winner and the loser, and that the first takes half of this total; that is to say one adds 2 to 8, which will be 10, of which half, 5, belongs to the first.

The First Formal Test of Significance

Comments on Arbuthnott (1710)

1   Introduction

In the years around 1700 the "argument from design" for the existence of God emerged from the mists of the classical past to become, in the hands of John Arbuthnott, a probability calculation involving the rejection of a null hypothesis on the grounds of the small probability of the observed data given that hypothesis. (The clarifying terminology "null hypothesis" was not coined until 1935, by R.A. Fisher.) The evolution of the argument took place among a small group of Fellows of the Royal Society of London, including Richard Bentley, Abraham de Moivre, Isaac Newton, Samuel Clarke, and William Derham, as well as Arbuthnott himself, leading Hacking (1975, quoting Anders Jeffner) to dub it "Royal Society theology." By 1718 de Moivre was stating its basis clearly in the Preface to the first edition of *The Doctrine of Chances*:

> Further, The same Arguments which explode the notion of Luck may, on the other side, be useful in some Cases to establish a due comparison between Chance and Design: We may imagine Chance and Design to be as it were in Competition with each other, for the production of some sorts of Events, and may calculate what Probability there is, that those Events should be rather owing to one than to the other.

Insofar as the argument was used to prove the existence of God from "the Appearances of Nature" it reached its apotheosis in Paley's *Evidences* (1802), the staple diet of generations of Cambridge University students, including Charles Darwin. It was Darwin, of course, who demolished it by discovering a mechanism — natural selection acting on heritable variation — which provided an alternative explanation. In an aphorism attributed to Fisher, "Natural selection is a mechanism for generating an exceedingly high degree of improbability."

The Reverend Richard Bentley (1662–1742; from 1700 Master of Trinity College, Cambridge) had made the argument explicit in his Boyle Lectures. In the eighth, delivered on December 5, 1692 (Bentley, 1693), he argued that it would be extremely improbable for all the planets to have ended up orbiting the sun in the same direction by chance, and almost in the same plane:

> Now, if the system had been fortuitously formed by the conven-
> ing matter of a chaos, how is it conceivable that all the planets,
> both primary and secondary [i.e., satellites], should revolve the
> same way, from west to east, and that in the same plane too,
> without any considerable variation? No natural and necessary
> cause could so determine their motions; and 'tis millions of mil-
> lions odds to an unit in such a cast of a chance. Such an apt
> and regular harmony, such an admirable order and beauty, must
> deservedly be ascribed to divine art and conduct . . . .

During the preparation of his lectures Bentley had applied to Newton for help on several points (Westfall, 1980). Newton's first letter in reply, dated December 10, 1692, opens "Sir, When I wrote my treatise about our system [*Principia*, 1687], I had an eye upon such principles as might work with considering men for the belief of a Deity. . . ." Later in the letter Newton writes:

> To your second query, I answer, that the motions which the
> planets now have could not spring from any natural cause alone,
> but were impressed by an intelligent Agent. For since comets
> descend into the region of our planets, and here move all man-
> ner of ways, going sometimes the same way with the planets,
> sometimes the contrary way, and sometimes in cross ways, in
> planes inclined to the plane of the ecliptic, and at all kinds of
> angles, 'tis plain that there is no natural cause which could de-
> termine all the planets, both primary and secondary, to move
> the same way and in the same plane, without any considerable
> variation: this must have been the effect of counsel.

The harmony in the system, concluded Newton, "was the effect of choice rather than chance" (Dyce, 1838).

Newton introduced Bentley's example into the second edition of his *Opticks*, which appeared in Latin (*Optice*) in 1706, translated by Samuel Clarke. Query 23 (in the subsequent English edition of 1717) contains the following.

> For while Comets move in very excentrick Orbs in all manner
> of Positions, blind Fate could never make all the Planets move
> one and the same way in Orbs concentrick, some inconsiderable
> Irregularities excepted, which may have arisen from the mutual
> Actions of Comets and Planets upon one another, and which
> will be apt to increase, till this System wants a Reformation.
> Such a wonderful Uniformity in the Planetary System must be
> allowed the Effect of Choice.

From this point on, with Newton's imprimatur, the argument flourished, leading to its use by other Fellows of the Royal Society such as Derham, de Moivre, and, of course, Arbuthnott himself, to whose paper we now turn (read to the Society in 1711 but published in "1710," a reflection of the vagaries of publishing at that time).

2   Arbuthnott's Calculation

John Arbuthnott, physician to Queen Anne, friend of Jonathan Swift and Isaac Newton, and creator of the character John Bull, was no stranger to probability (Aitken, 1892). In 1692 he had published (anonymously) a translation of Huygens's *De ratiociniis in ludo aleæ* (1657) as *Of the Laws of Chance*, adding "I believe the Calculation of the Quantity of Probability might be improved to a very useful and pleasant Speculation, and applied to a great many events which are accidental, besides those of Games." There exists a 1694 manuscript of Arbuthnott's which foreshadows his 1710 paper (Bellhouse, 1989), and in 1701, again anonymously, he published an *Essay on the Usefulness of Mathematical Learning* in which he argued that a thorough understanding of the "Theory of *Chance*" "requires a great knowledge of Numbers, and a pretty competent one of *Algebra*."

Arbuthnott's paper is only five pages long. There are several excellent commentaries on it (Hacking, 1965, 1975; Hald, 1990; Shoesmith, 1987), but we are here primarily concerned with just one paragraph, the solution to the *Problem* on page 188: "A lays against B, that every Year there shall be born more Males than Females: to find A's Lot, or the Value of his Expectation." The argument runs as follows.

Assume, as earlier in the paper, that the probability of a male birth is 1/2. Then in any year the probability that there are more males than females born is less than 1/2 because that event excludes exact equality which, by what has gone before, has a small but nonzero probability (actually only for an even number of births, as Arbuthnott appreciates). Let us therefore assume that the probability is exactly 1/2 "[so] that the Argument may be stronger." For there to be more males 82 years in succession the probability is then $(1/2)^{82}$, or $1/(4.836 \times 10^{24})$, and for this improbable event to be repeated in all ages, and all over the world, the probability is "near an infinitely small Quantity." "From whence it follows, that it is Art, not Chance, that governs." We note that since the event discussed corresponds to the extreme tail of a binomial distribution the question does not arise as to whether even more extreme or even less probable events should be included in the argument: there are none. Moreover, Arbuthnott's test is "one-sided," for he is explicitly considering only an excess of males.

One puzzle remains about this argument. No commentator has touched on the additional reason advanced by Arbuthnott for "A's chance" to be even smaller than calculated: "but that this Excess shall happen in a constant proportion, and the Difference lye within fix'd limits." This says that the event on which A bets is no longer to be a simple excess of males over females in each and every year, but an excess *within certain fixed limits*. Such an event would indeed have a smaller probability than a simple excess, thus strengthening Arbuthnott's argument just as he asserts.

The "constant proportion" is mentioned in Arbuthnott's previous paragraph, and is none other than the excess of males over females which "appears from the annexed Table." Nicholas Bernoulli later used the value $p = 18/35$ for the proportion of male births in this case (see Hald, 1990), and thus an example of "fix'd limits" for the results might be $17/35$ and $19/35$. The probability of an annual result within these limits is of course much smaller than the $1/2$ used by Arbuthnott in his test, but what he has not noticed is the associated *reductio ad absurdum*, for the natural extension of this argument would be to compute the probability of the actual data exactly, which would not only be extremely small, but *would be extremely small on any hypothesis as to the value of p whatever. Every* hypothesis about p would have to be rejected. As Hacking (1965) observes, the resolution of this problem leads in the direction of likelihood ratios. Arbuthnott, however, is naturally blind to these concerns, and in any case might not have been able to compute the probabilities on any alternative hypothesis such as $p = 18/35$ since the binomial distribution for general p was not published until 1713 (Montmort; James Bernoulli; see Edwards, 1987).

3   The Sequels to Arbuthnott's Paper

Arbuthnott's argument was used by "Royal Society theologians" such as Derham and de Moivre (who added a mention of it to the third edition of *The Doctrine of Chances*, 1756), and on the continent his statistical analysis was elaborated by Nicholas Bernoulli and Willem 'sGravesande. Bernoulli had visited London in 1712 and had subsequently met 'sGravesande in Holland. All this work was later to become well known through its inclusion in Todhunter's *History* (1865). We refer the reader to Shoesmith (1985, 1987) and Hald (1990) for the details, and here note only that 'sGravesande made precisely the computation envisaged by Arbuthnott, using limits of $5745/11{,}429$ and $6128/11{,}429$.

The question of the variability in Arbuthnott's table, initially taken up by Bernoulli, was the subject of a $\bar{\chi}^2$-test by Anscombe (1981) who found $\bar{\chi}^2 = 170$ with 81 degrees of freedom, a result confirmed by Hald (1990) using standardized normal deviates.

In 1734 Daniel Bernoulli, Nicholas's first cousin, took up Bentley's original point about the inclination of the planetary orbits in response to a prize question proposed by the Academy of Sciences of Paris. He extended Arbuthnott's method of computing probabilities on the basis of the null hypothesis of chance by computing the probability of the observed result or *one more extreme*, thus introducing the modern notion of a rejection region. Todhunter (1865) gives a critical analysis.

Finally, a statistical audience might like to be reminded that the evolutionary reason for the slight excess of male births noted by Arbuthnott was first given by Fisher (1930) in *The Genetical Theory of Natural Selection*, in a section famous in evolutionary biology for its game-theoretic basis. The underlying argument was originally due to Charles Darwin (see Edwards, 1998).

4   Postscript

We leave the last word to Arbuthnott himself. The year before he died in 1735 he published "A Poem: Know Yourself" (Aitken, 1892), echoing Newton's phrase "blind fate" in the opening lines:

> *What am I? how produced? and for what end?*
> *Whence drew I being? to what period tend?*
> *Am I the abandoned orphan of blind chance,*
> *Dropt by wild atoms in disordered dance?*
> *Or from an endless chain of causes wrought?*
> *And of unthinking substance, born with thought?*

References

Aitken, G.A. (1892). *The Life and Works of John Arbuthnot.* Clarendon Press, Oxford.

Anscombe, F.J. (1981). *Computing in Statistical Science through APL.* Springer, New York.

Arbuthnott, J. (1710). An argument for Divine Providence, taken from the constant regularity observ'd in the births of both sexes. *Phil. Trans.,* **27**, 186–190.

Bellhouse, D.R. (1989). A manuscript on chance written by John Arbuthnot. *Intern. Statist. Rev.* **57**, 249–259.

Bentley, R. (1693). *The Folly and Unreasonableness of Atheism.* Reprinted in Dyce (1838).

Dyce, A. (ed.) (1838). *The Works of Richard Bentley D.D.*, Vol. 3. Macpherson, London.

Edwards, A.W.F. (1987). *Pascal's Arithmetical Triangle.* Griffin, London and Oxford University Press, New York.

Edwards, A.W.F. (1998). Natural selection and the sex ratio: Fisher's sources. *American Naturalist*, **151**, 564–569.

Fisher, R.A. (1930). *The Genetical Theory of Natural Selection*. Clarendon Press, Oxford; 2nd edn. Dover, New York, 1958; variorum edn. Oxford University Press, 1999.

Fisher, R.A. (1935). *The Design of Experiments*. Oliver & Boyd, Edinburgh.

Hacking, I. (1965). *Logic of Statistical Inference*. Cambridge University Press.

Hacking, I. (1975). *The Emergence of Probability*. Cambridge University Press.

Hald, A. (1990). *A History of Probability and Statistics and Their Applications before 1750*. Wiley, New York.

Moivre, A. de (1718). *The Doctrine of Chances*. Pearson, London.

Paley, W. (1802). *Natural Theology — or Evidences of the Existence and Attributes of the Deity Collected from the Appearances of Nature*. Vincent, Oxford.

Shoesmith, E. (1985). Nicholas Bernoulli and the argument for Divine Providence. *Intern. Statist. Rev.*, **53**, 255–259.

Shoesmith, E. (1987). The continental controversy over Arbuthnot's argument for Divine Providence. *Historia Mathematica*, **14**, 133–146.

Todhunter, I. (1865). *A History of the Mathematical Theory of Probability*. Macmillan, Cambridge. [Reprinted by Chelsea, New York, 1949, 1965.]

Westfall, R.S. (1980). *Never at Rest — A Biography of Newton*. Cambridge University Press.

(186)

II. *An Argument for Divine Providence, taken from the constant Regularity observ'd in the Births of both Sexes. By Dr. John Arbuthnott, Physitian in Ordinary to Her Majesty, and Fellow of the College of Physitians and the Royal Society.*

Among innumerable Footsteps of Divine Providence to be found in the Works of Nature, there is a very remarkable one to be observed in the exact Ballance that is maintained, between the Numbers of Men and Women ; for by this means it is provided, that the Species may never fail, nor perish, since every Male may have its Female, and of a proportionable Age. This Equality of Males and Females is not the Effect of Chance but Divine Providence, working for a good End, which I thus demonstrate :

Let there be a Die of Two sides, M and F, (which denote Cross and Pile), now to find all the Chances of any determinate Number of such Dice, let the Binome $M+F$ be raised to the Power, whose Exponent is the Number of Dice given ; the Coefficients of the Terms will shew all the Chances sought. For Example, in Two Dice of Two sides $M+F$ the Chances are $M^2 + 2MF + F^2$, that is, One Chance for M double, One for F double, and Two for M single and F single ; in Four such Dice there are Chances $M^4 + 4M^3F + 6M^2F^2 + 4MF^3 + F^4$, that is, One Chance for M quadruple, One for F quadruple, Four for triple M and single F, Four for single M and triple F, and Six for M double and F double ; and universally, if the Number of Dice be n, all their Chances will be expressed in this Series

$$M^n +$$

$$M^0 + \tfrac{n}{1} \times M^{n-1}F + \tfrac{n}{1} \times \tfrac{n-1}{2} \times M^{n-2}F^2 + \tfrac{n}{1} \times \tfrac{n-1}{2} \times \tfrac{n-2}{3} \times M^{n-3}F^3 +, \; \&c.$$

It appears plainly, that when the Number of Dice is even there are as many M's as F's in the middle Term of this Series, and in all the other Terms there are moſt M's or moſt F's.

If therefore a Man undertake with an even Number of Dice to throw as many M's as F's, he has all the Terms but the middle Term againſt him ; and his Lot is to the Sum of all the Chances, as the coefficient of the middle Term is to the power of 2 raiſed to an exponent equal to the Number of Dice : ſo in Two Dice his Lot is $\tfrac{2}{4}$ or $\tfrac{1}{2}$, in Three Dice $\tfrac{6}{16}$ or $\tfrac{3}{8}$, in Six Dice $\tfrac{20}{64}$ or $\tfrac{5}{16}$, in Eight $\tfrac{70}{256}$ or $\tfrac{35}{128}$, $\&c.$

To find this middle Term in any given Power or Number of Dice, continue the Series $\tfrac{n}{1} \times \tfrac{n-1}{2} \times \tfrac{n-2}{3}$, $\&c.$ till the number of terms are equal to $\tfrac{1}{2}n$. For Example, the coefficient of the middle Term of the tenth Power is $\tfrac{10}{1} \times \tfrac{9}{2} \times \tfrac{8}{3} \times \tfrac{7}{4} \times \tfrac{6}{5} = 252$, the tenth Power uf 2 is 1024, if therefore A undertakes to throw with Ten Dice in one throw an equal Number of M's and F's, he has 252 Chances out of 1024 for him, that is his Lot is $\tfrac{252}{1024}$ or $\tfrac{63}{256}$, which is leſs than $\tfrac{1}{4}$.

It will be eaſy by the help of Logarithms, to extend this Calculation to a very great Number, but that is not my preſent Deſign. It is viſible from what has been ſaid, that with a very great Number of Dice, A's Lot would become very ſmall ; and conſequently (ſuppoſing M to denote Male and F Female) that in the vaſt Number of Mortals, there would be but a ſmall part of all the poſſible Chances, ıor its happening at any aſſignable time, that an equal Number of Males and Females ſhould be born.

It is indeed to be confeſſed that this Equality of Males and Females is not Mathematical but Phyſical, which alters much the foregoing Calculation ; for in this Caſe
                                                                    the

(188)

the middle Term will not exactly give A's Chances, but his Chances will take in some of the Terms next the middle one, and will lean to one side or the other. But it is very improbable (if mere Chance govern'd) that they would never reach as far as the Extremities: But this Event is wisely prevented by the wise Oeconomy of Nature; and to judge of the wisdom of the Contrivance, we must observe that the external Accidents to which Males are subject (who must seek their Food with danger) do make a great havock of them, and that this loss exceeds far that of the other Sex, occasioned by Diseases incident to it, as Experience convinces us. To repair that Loss, provident Nature, by the Disposal of its wise Creator, brings forth more Males than Females; and that in almost a constant proportion. This appears from the annexed Tables, which contain Observations for 82 Years of the Births in *London*. Now, to reduce the Whole to a Calculation, I propose this

    Problem. A lays against B, that every Year there shall be born more Males than Females: To find A's Lot, or the Value of his Expectation.

    It is evident from what has been said, that A's Lot for each Year is less than $\frac{1}{2}$; (but that the Argument may be stronger) let his Lot be equal to $\frac{1}{2}$ for one Year. If he undertakes to do the same thing 82 times running, his Lot will be $\overline{\frac{1}{2}}|^{82}$, which will be found easily by the Table of Logarithms to be $\dfrac{1}{4\ 8360\ 0000\ 0000\ 0000\ 0000\ 0000}$. But if A wager with B, not only that the Number of Males shall exceed that of Females, every Year, but that this Excess shall happen in a constant Proportion, and the Difference lye within fix'd limits; and this not only for 82 Years, but for Ages of Ages, and not only at *London*, but all over the World; (which 'tis highly probable is Fact, and designed that every Male may have a Female of the same Country and suitable Age) then A's Chance will be near an infinitely small Quantity, at least

less

(189)

lefs than any affignable Fraction. From whence it follows, that it is Art, not Chance, that governs.

There feems no more probable Caufe to be affigned in
Phyficks for this Equality of the Births, than that in
our firft Parents Seed there were at firft formed an equal
Number of both Sexes.

Scholium. From hence it follows, that Polygamy is
contrary to the Law of Nature and Juftice, and to the
Propagation of Human Race; for where Males and
and Females are in equal number, if one Man takes
Twenty Wives, Nineteen Men muft live in Celibacy,
which is repugnant to the Defign of Nature; nor is it
probable that Twenty Women will be fo well impreg-
nated by one Man as by Twenty.

Chriftened.			Chriftened.		
Anno.	*Males*	*Females.*	*Anno.*	*Males.*	*Females.*
1629	5218	4683	1648	3363	3181
30	4858	4457	49	3079	2746
31	4422	4102	50	2890	2722
32	4994	4590	51	3231	2840
33	5158	4839	52	3220	2908
34	5035	4820	53	3196	2959
35	5106	4928	54	3441	3179
36	4917	4605	55	3655	3349
37	4703	4457	56	3668	3382
38	5359	4952	57	3396	3289
39	5366	4784	58	3157	3013
40	5518	5332	59	3209	2781
41	5470	5200	60	3724	3247
42	5460	4910	61	4748	4107
43	4793	4617	62	5216	4803
44	4107	3997	63	5411	4881
45	4047	3919	64	6041	5681
46	3768	3395	65	5114	4858
47	3796	3536	66	4678	4319

B b

Chriftened.

(190)

Anno.	Males.	Females.	Anno.	Males.	Females.
1667	5616	5322	1689	7604	7167
68	6073	5560	90	7909	7302
69	6506	5829	91	7662	7392
70	6278	5719	92	7602	7316
71	6449	6061	93	7676	7483
72	6443	6120	94	6985	6647
73	6073	5822	95	7263	6713
74	6113	5738	96	7632	7229
75	6058	5717	97	8062	7767
76	6552	5847	98	8426	7626
77	6423	6203	99	7911	7452
78	6568	6033	1700	7578	7061
79	6247	6041	1701	8102	7514
80	6548	6299	1702	8031	7656
81	6822	6533	1703	7765	7683
82	6909	6744	1704	6113	5738
83	7577	7158	1705	8366	7779
84	7575	7127	1706	7952	7417
85	7484	7246	1707	8379	7687
86	7575	7119	1708	8239	7623
87	7737	7214	1709	7840	7380
88	7487	7101	1710	7640	7288

Coincidences and the Method of Inclusion and Exclusion

Comments on Montmort (1713), N. Bernoulli (1713), and de Moivre (1718)

1 Introduction

The most familiar form of the problem of coincidences or matches goes back to Montmort (1708) and may be stated as follows: successive drawings without replacement are made from randomly shuffled tickets numbered $1, 2, \ldots, n$. What is the probability of at least one match (i.e., ticket i comes up at the ith draw, $i = 1, \ldots, n$)?

A convenient modern way of solving this problem is through the method of inclusion and exclusion, by which for events $A_1, A_2, \ldots, A_n$,

$$P\left(\bigcup_{i=1}^{n} A_i\right) = \sum_{r=1}^{n}(-1)^{r-1} \sum P(A_{i_1} A_{i_2} \ldots A_{i_r}), \tag{a}$$

where the inner sum ranges over all integers $i_1, i_2, \ldots, i_r$ such that $1 \leq i_1 < i_2 < \cdots < i_r \leq n$. In the important special case where $A_1, A_2, \ldots, A_n$ are exchangeable, eq. (a) becomes

$$P\left(\bigcup_{i=1}^{n} A_i\right) = \sum_{r=1}^{n}(-1)^{r-1} \binom{n}{r} P(A_1 A_2 \ldots A_r). \tag{b}$$

Now let A_i represent a match at the ith draw. Then $P(A_1 A_2 \ldots A_r) = \frac{1}{n} \cdot \frac{1}{n-1} \cdots \frac{1}{n-r+1}$, so that from (b) the probability, $P_1^{(n)}$, of at least one match is

$$P_1^{(n)} = \sum_{r=1}^{n}(-1)^{r-1}\frac{1}{r!} = 1 - \frac{1}{2!} + \frac{1}{3!} - \cdots (-1)^{n+1}\frac{1}{n!}, \tag{c}$$

which quickly approaches $1 - e^{-1} = 0.63212\ldots$.

The historical development of this subject has been admirably presented by Hald (1990, Chapter 19). Here we feature the pioneering articles by Montmort and de Moivre as well as part of a letter by Nicholas Bernoulli to Montmort in which he provides arguments to justify results stated without proof by Montmort. Interest was not confined to obtaining $P_1^{(n)}$ but included the probability $P_{1,i}^{(n)}$ that the first match occurs with the ith card. Montmort wrote in French and Bernoulli in Latin. De Moivre, fortunately writing in English, saw the general features of the problem of coincidences. We reproduce here Problem XXV of de Moivre (1718).

2   Montmort and Nicholas Bernoulli

After explaining the Game of Thirteen, Montmort considers the simpler situation described above, determines $P_{1,i}^{(n)}$ successively for $n = 2, 3, 4, 5$ and $i = 1, \ldots, n$, and obtains $P_1^{(n)}$ as $\sum_{i=1}^{n} P_{1,i}^{(n)}$. He appears to proceed by enumeration of all possible arrangements of n letters. Then in his Section 104 he takes a big jump to state a recursion result, which in our notation is

$$P_1^{(n)} = \left[(n-1) P_1^{(n-1)} + P_1^{(n-2)} \right] / n. \tag{d}$$

It is easily verified that (d) follows from (c), but it is not clear how Montmort obtained (d). Bernoulli gives a proof in his letter, as described by Hald.

Montmort uses another recursion formula in the construction of the interesting first table in his Section 106. Let $N_i^{(n)} = n! P_{1,i}^{(n)}$, the number of arrangements of n cards such that the first match occurs with the ith card. The entries in the nth row of the table are $N_n^{(n)}, N_{n-1}^{(n)}, \ldots, N_1^{(n)}$ and, following the equality sign, their sum $N^{(n)}$. Montmort's construction formula is, for $i = 1, \ldots, n$,

$$N_{i+1}^{(n)} = N_i^{(n)} - N_i^{(n-1)} , \tag{e}$$

where $N_1^{(n)} = (n-1)!$. Montmort's argument is only hinted at, but perhaps went as follows: the number of arrangements of n cards with the first match at the $(i+1)$th card, $N_{i+1}^{(n)}$, must be the same as the number with the first match at the ith card, $N_i^{(n)}$, except for the possibility that with the $(i+1)$th card in position $i+1$, the first match already occurred with the ith card. We must therefore subtract from $N_i^{(n)}$ the number of arrangements, $N_i^{(n-1)}$, of the $n-1$ free cards for which the first match occurs at the ith card. Note that matches earlier than with the ith card have already been allowed for in the calculation of $N_i^{(n)}$.

In the part of his letter presented here Bernoulli seems to use a somewhat different argument by applying the principle of inclusion and exclusion directly to the calculation of $N_i^{(n)}$. For example, to find $N_3^{(n)}$ he appears to reason that $N_3^{(n)}$ equals $N_1^{(n)}$, except that, given a match in position 3, we must allow for first coincidences in positions 1 or 2. This means we must subtract $2(n-2)!$, except that then we have double-counted the outcome with a match in both positions 1 and 2. Hence, we must add $(n-3)!$, and

so on. As can be seen, Bernoulli quickly arrives at eq. (c). But see also
Todhunter (1865, Section 161).

3   De Moivre and Some Generalizations

Although in his Problem XXV, de Moivre makes no reference to Montmort
or Nicholas Bernoulli, he makes it clear in the Preface of *The Doctrine
of Chances* that he is familiar with their work on the problem of coincidences. De Moivre is evidently pleased with his own approach, consisting of
inclusion-exclusion arguments applied to a large class of cases that include
eq. (c). He writes:

> *In the 24th and 25th Problems, I explain a new sort of Algebra,
> whereby some Questions relating to Combinations are solved by
> so easy-a Process, that their solution is made in some measure
> an immediate consequence of the Method of Notation. I will not
> pretend to say that this new Algebra is absolutely necessary to
> the Solving of those Questions which I make to depend on it,
> since it appears by Mr.* De Montmort's *Book, that both he and
> Mr.* Nicholas Bernoully [sic] *have solved, by another Method,
> many of the cases therein proposed: But I hope I shall not be
> thought guilty of too much Confidence, if I assure the Reader,
> that the Method I have followed has a degree of Simplicity, not
> to say of Generality, which will hardly be attained by any other
> Steps than by those I have taken.*

De Moivre's results are stated with unusual clarity. Their justification is
not quite as clear, but he develops an interesting operational method (see
Hald, 1990, p. 336), that leads to eq. (b) and even its generalization to the
probability of occurrence of at least m exchangeable events, $P_m^{(n)}$:

$$P_m^{(n)} = \sum_{r=m}^{n} (-1)^{r-m} \binom{r-1}{m-1} \binom{n}{r} P(A_1 A_2 \ldots A_r). \tag{f}$$

Hald goes on to give an interesting account of the subsequent history of
the problem of coincidences. We note only one of his many references. An
important generalization of (f) to general events $A_1, \ldots, A_n$ was made in
an elegant short article by Jordan (1867). This amounts simply to replacing

$$\binom{n}{r} P(A_1 \ldots A_r) \quad \text{by} \quad \sum P(A_{i_1} \ldots A_{i_r}) ,$$

where the sum is as in (a). Jordan's argument is completely self-contained,
with no reference to earlier work.

4 Applications to Statistics

An important class of statistical applications results if A_i is taken to be $X_i > x$, $i = 1, \ldots, n$, where the X_i are any random variables. Then (a) becomes

$$P(X_{n:n} > x) = \sum_{r=1}^{n} (-1)^{r-1} \sum P(X_{i_1} > x, \ldots, X_{i_r} > x), \qquad \text{(g)}$$

where $X_{n:n} = \max(X_1, \ldots, X_n)$. If $X_1, \ldots, X_n$ are exchangeable variates, then

$$P(X_{n:n} > x) = \sum_{r=1}^{n} (-1)^{r-1} \binom{n}{r} P(X_1 > x, \ldots, X_r > x). \qquad \text{(h)}$$

The first use of (h) appears to have been made by Cochran (1941) in obtaining upper 5% points of the statistic

$$T = Y_{n:n} \bigg/ \left(\sum_{i=1}^{n} Y_i \right),$$

where the Y_i are independent χ^2 variates with ν degrees of freedom. The statistic provides a test of homogeneity of variance when n independent normal samples of size $\nu+1$ are available. Large values of T lead to rejection of the homogeneity assumption. To apply (h), we set $X_i = Y_i/S$, where $S = \sum_{j=1}^{n} Y_j$. Note that if $x > \frac{1}{2}$, then $P(X_i > x, X_j > x) = 0$, $i \neq j$, so that (h) then reduces to $P(X_{n:n} > x) = nP(X_1 > x)$, which is easily found.

More generally, if $X_1, \ldots, X_n$ are negatively dependent, the terms in (g) and (h) decrease with r for x large. Then the first term ($r = 1$), or the first two, often give a good approximation, or can be usefully bounded. For many other statistical applications of the principle of inclusion and exclusion, see David (1981, Sections 5.3, 5.4, 5.6, and 8.4) and Galambos and Simonelli (1996, Chapters 8 and 9).

References

Cochran, W.G. (1941). The distribution of the largest of a set of estimated variances as a fraction of their total. *Ann. Eugen.*, **11**, 47–52.

David, H.A. (1981). *Order Statistics*, 2nd edn. Wiley, New York.

Galambos, J. and Simonelli, I. (1996). *Bonferroni-type Inequalities with Applications.* Springer, New York.

Hald, A. (1990). *A History of Probability and Statistics and Their Applications before 1750.* Wiley, New York.

Jordan, C. (1867). De quelques formules de probabilité. *C. R. Acad. Sci.*, Paris, **65**, 993–994.

Moivre, A. de (1718). *The Doctrine of Chances: or, A Method of Calculating the Probabilities of Events in Play*. Pearson, London.

Moivre, A. de (1756). *The Doctrine of Chances*, 3rd edn. Millar, London. [Reprinted by Chelsea, New York, 1967]

Montmort, P.R. de (1708). *Essay d'analyse sur les jeux de hazard*. Quillau, Paris.

Montmort, P.R. de (1713). *Essay d'analyse sur les jeux de hazard*, 2nd edn. *Revûe et augmentée de plusieurs lettres*. Quillau, Paris. [Reprinted by Chelsea, New York, 1980]

Todhunter, I. (1865). *A History of the Mathematical Theory of Probability*. Macmillan, London. [Reprinted by Chelsea, New York, 1949]

On the Game of Thirteen

P. R. de Montmort

Explanation of the Game

98. The players first draw a card to determine the banker. Let us suppose that this is Peter and that the number of players is as desired. From a complete pack of fifty-two cards, judged adequately shuffled, Peter draws one after the other, calling one as he draws the first card, two as he draws the second, three as he draws the third, and so on, until the thirteenth, calling king. Then, if in this entire sequence of cards he has not drawn any with the rank he has called, he pays what each of the players has staked and yields to the player on his right.

But if in the sequence of thirteen cards, he happens to draw the card he calls, for example, drawing an ace as he calls one, or a two as he calls two, or a three as he calls three, and so on, then he takes all the stakes and begins again as before, calling one, then two, and so on.

It may happen that after having won several times and recommenced with one, Peter does not have enough cards in his hand to go up to thirteen. Then, lacking the pack, he must shuffle the cards, let them be cut, and draw from the complete pack the number of cards he needs to continue the game. He must begin where he left off with the preceding hand. For example, if he called seven when drawing the last card, then upon drawing the first card from the complete pack, after it has been cut, he must call eight and then nine, and so on, up to thirteen. He stops there unless he wins again, in which case he recommences, first calling one, then two, and the rest, as just explained. From which it appears that Peter can play several hands in succession, and even that he can continue the game indefinitely.

Problem
Proposition V

Peter has a certain number of different cards that are not re- peated and that are shuffled as desired. He wagers Paul that if he draws them in sequence and calls them according to the order of the cards, beginning either from the highest or the lowest, he will at least once happen to draw the card he calls. For exam- ple, with a hand of four cards, namely an ace, a two, a three, and a four, shuffled as desired. Peter bets that upon drawing them in sequence and calling one as he draws the first, two as he draws the second, three as he draws the third, he will hap- pen to draw either an ace as he calls one, or draw a two as he

*calls two, or draw a three as he calls three, or draw a four as
he calls four. Suppose the same holds for any other number of
cards. Required: Peter's chance or expectation, for any number
of cards, from two up to thirteen.*

99. Let the cards with which Peter plays be represented by the letters
a, b, c, d, etc. If m denotes the number of cards he holds and n the
number of all possible arrangements of these cards, then the fraction n/m
will represent how many different times each letter will occupy each place.
Now it must be noted that these letters do not always find their place in a
manner useful to the banker. For example, a, b, c produces only one win
to the person with the cards although each of these three letters is in its
place. Likewise, b, a, c, d produces only one win for Peter, although each
of the letters c and d is in its place. The difficulty of this problem consists
in disentangling how many times each letter is in its place usefully for Peter
and how many times it is useless.

First Case

*Peter holds an ace and a two, and has shuffled the two cards.
He wagers Paul that, calling one when he draws the first card
and two when he [draws] the second, he will succeed in either
drawing an ace as the first card or in drawing a two for the
second card. The money in the game is denoted by A.*

100. Two cards can arrange themselves only in two different ways: the one
makes Peter win, the other makes him lose. Hence, his expectation will be
$\frac{A+0}{2} = \frac{1}{2}A$.

Second Case

Peter holds three cards.

101. Let these three cards be represented by the letters a, b, c. We observe
that, of the six different arrangements possible for these three letters, there
are two with a in first place; there is one with b in second place, a not
having been in first place; and one where c is in third place, a not having
been in first place and b not having been in second place. It follows that
$S = \frac{2}{3}A$ and consequently that Peter's expectation is to Paul's as two is to
one.

Third Case

Peter holds four cards.

102. Let the four cards be represented by the letters a, b, c, d. We observe
that, of the twenty-four different arrangements possible with these four
letters, there are six with a occupying the first place; there are four with

b second, a not being first; three with c third, a not being first and b not being second; finally, two with d fourth, a not being first, b not being second, and c not being third. It follows that Peter's expectation $= S = \frac{6+4+3+2}{24}A = \frac{15}{24}A = \frac{5}{8}A$. Consequently Peter's expectation is to Paul's as five is to three.

Fourth Case

Peter holds five cards.

103. Let the five cards be represented by the letters a, b, c, d, f. We observe that of the 120 different arrangements possible with five letters, there are twenty-four with a occupying the first place, eighteen with b occupying the second place, a not being first; fourteen with c in third place, a not being in first place nor b in second; eleven with d in fourth place, a not being first, nor b second, nor c third; finally, there are nine arrangements with f in fifth place, a not being in first place, nor b in second, nor c in third, nor d in fourth. It follows that Peter's expectation $= S = \frac{24+18+14+11+9}{120}A = \frac{76}{120}A = \frac{19}{30}A$ and consequently that Peter's expectation is to Paul's as nineteen is to eleven.

Generally

104. Let S denote the desired expectation when Peter holds p cards; g, Peter's expectation when there are $p-1$ cards; d, his expectation when he holds $p-2$ cards. Then we will have $S = [g(p-1) + d]/p$. This formula covers all cases and we can see the result in the following table.

Table

If $p = 1$, we have $S = A$.

If $p - 2$, we have $S - \frac{1}{2}A$.

If $p = 3$, we have $S = \frac{2}{3}A = \frac{1}{2}A + \frac{1}{6}A$.

If $p = 4$, we have $S = \frac{5}{8}A = \frac{1}{2}A + \frac{1}{8}A$.

If $p = 5$, we have $S - \frac{19}{30}A - \frac{1}{2}A + \frac{2}{15}A$.

If $p = 6$, we have $S = \frac{91}{144}A = \frac{1}{2}A + \frac{19}{144}A$.

If $p = 7$, we have $S = \frac{531}{840}A = \frac{1}{2}A + \frac{111}{840}A$.

If $p = 8$, we have $S = \frac{3641}{5760}A = \frac{1}{2}A + \frac{761}{5760}A$.

If $p = 9$, we have $S = \frac{28673}{45360}A = \frac{1}{2}A + \frac{5993}{45360}A$.

If $p = 10$, we have $S = \frac{28319}{44800}A = \frac{1}{2}A + \frac{5919}{44800}A$.

If $p = 11$, we have $S = \frac{2523223}{3991680}A = \frac{1}{2}A + \frac{527383}{3991680}A$.

If $p = 12$, we have $S = \frac{302786759}{479001600}A = \frac{1}{2}A + \frac{63285959}{479001600}A$.

If $p = 13$, we have $S = \frac{109339663}{172972800}A = \frac{1}{2}A + \frac{22853263}{172972800}A$.

Likewise, this formula gives Peter's expectation if we suppose that there is a larger number of cards of different kinds.

Remark I

105. The preceding solution provides a remarkable use of the numbers involved. For, I find, on examining the formula, that Peter's expectation is expressible as an infinite sequence of terms that are alternately positive and negative, such that the numerator is the sequence of numbers in the Table of Section 1 making up the vertical column corresponding to p and beginning with p. The denominator is the sequence of products $p(p-1)(p-2)(p-3)(p-4)(p-5)$, etc., such that upon canceling those products appearing in both numerator and denominator, there remains the following very simple expression for Peter's chance: $\frac{1}{1} - \frac{1}{1.2} + \frac{1}{1.2.3} - \frac{1}{1.2.3.4} + \frac{1}{1.2.3.4.5} - \frac{1}{1.2.3.4.5.6} + \cdots$.

Remark II

106. The two formulae of Sections 104 and 105 tell us how many chances the person with the cards has in order to win with some card or other. But they do not let us know how many chances he has for each card he draws, from the first to the last. It is obvious that this number of chances always decreases and that there are, for example, more chances to win with the ace than with two, and with three than with four, etc. But it is not easy to determine from the above the law of this decrease. It can be found in the following table.

$$
\begin{array}{rrrrrrr}
1 = 1 \\
0. & 1 = 1 \\
1. & 1. & 2 = 4 \\
2. & 3. & 4. & 6 = 15 \\
9. & 11. & 14. & 18. & 24 = 76 \\
44. & 53. & 64. & 78. & 96. & 120 = 455 \\
265. & 309. & 362. & 426. & 504. & 600. & 720 = 3186 \\
1854. & 2119. & 2428. & 2790. & 3216. & 3720. & 4320. & 5040 = 25487.
\end{array}
$$

This table shows that with five cards, for example, an ace, a two, a three, a four, and a five, Peter has twenty-four ways of winning with the ace; eighteen of winning with the two not having won with the ace; fourteen of winning with the three, not having won with either ace or two; eleven of winning with the four, not having won with the ace, nor the two, nor the three; and finally that he has only nine ways of winning with the five, not having won with the ace, nor the two, nor the three, nor the four.

Each row of this table is formed from the preceding row in a very simple way. To explain this, let us suppose again that there are five cards. We see at once that there are twenty-four ways of winning with the ace. This is obvious, since with the ace fixed in first place, the four other cards can be arranged in all possible ways. In general, it is clear that for p cards, the number of chances of winning with the ace can be expressed by as many products of the natural numbers 1, 2, 3, 4, 5, etc., as there are unities in

$p - 1$. With this established, $24 - 6 = 18$ gives me the chances of winning with the two, $18 - 4 = 14$ gives me the chances of winning with the three, $14 - 3 = 11$ gives me the chances of winning with the four, and finally $11 - 2 = 9$ gives me the chances of winning with the five.

The same holds for any other number of cards, and quite generally each number in the table is equal to the difference of the one on its right and the one immediately above it, which has already been found.

Moreover, we can find a regular pattern in the numbers $1, 1, 4, 15, 76, 455, \ldots$ that give all the ways of winning with any number of cards. This pattern can be seen in the following table:

$$
\begin{aligned}
\overline{0 \times 1} + 1 &= 1 \\
\overline{1 \times 2} - 1 &= 1 \\
\overline{1 \times 3} + 1 &= 4 \\
\overline{4 \times 4} - 1 &= 15 \\
\overline{15 \times 5} + 1 &= 76 \\
\overline{76 \times 6} - 1 &= 455 \\
455 \times 7 + 1 &= 3186 \\
3186 \times 8 - 1 &= 25487.
\end{aligned}
$$

These numbers, $1, 1, 4, 15, 76, \ldots$ show how many chances there are in order that any one among the p cards finds itself arranged in its place; that is to say, for example, the 3 in the 3rd place, or the 4 in the 4th, or the 5 in the 5th, etc.

Corollary I

107. Let p be the number of cards, g the number of chances that Peter has of winning when the number of cards is $p - 1$. The number of chances favorable to Peter is given by the very simple formula $pg \pm 1$; that is to say, $+$ when p is an odd number and $-$ when it is even.

Corollary II

108. The numbers $0, 1, 2, 9, 44, 265 \ldots$, which make up the first column of the table on the preceding page, give the number of chances there are that no card is in its place.

Letter from Nicholas Bernoulli to Montmort on the Game of Thirteen, reproduced in Montmort (1713, p. 301)

Let the cards held by Peter be denoted by the letters a, b, c, d, e, etc. and let their number be n. The number of all possible cases will be $= 1.2.3 \ldots n$; the number of cases with a in first place $= 1.2.3 \ldots n - 1$; the number of cases with b in second but a not in first place $= 1.2.3 \ldots n-1-1.2.3 \ldots n-2$; the number of cases with c in third place, with neither a in first nor b in second place $= 1.2.3 \ldots n-1-2 \times 1.2.3 \ldots n-2+1.2.3 \ldots n-3$; the number of cases with d in fourth, none of the preceding being in its correct place $= 1.2.3 \ldots n - 1 - 3 \times 1.2.3 \ldots n - 2 + 3 \times 1.2.3 \ldots n - 3 - 1.2.3 \ldots n - 4$. In general, the number of possible cases such that a card is correctly in mth place, but none of the preceding cards is in its place $= 1.2.3 \ldots n - 1 - \frac{m-1}{1} \times 1.2.3 \ldots n - 2 + \frac{m-1.m-2}{1.2} \times 1.2.3 \ldots n - 3 - \frac{m-1.m-2.m-3}{1.2.3} \times 1.2.3 \ldots n - 4 + \ldots$ up to $\pm \frac{m-1.m-2 \ldots m-m+1}{1.2.3 \ldots m-1} \times 1.2.3 \ldots n - m$. Hence the chance of the player who actually wishes to win with the mth card is $= \frac{1}{n} - \frac{m-1}{1} \times \frac{1}{n.n-1} + \frac{m-1.m-2}{1.2} \times \frac{1}{n.n-1.n-2} - \frac{m-1.m-2.m-3}{1.2.3} \times \frac{1}{n.n-1.n-2.n-3} + \ldots$ up to $\pm \frac{m-1.m-2 \ldots m-m+1}{1.2.3 \ldots m-1} \times \frac{1}{n.n-1 \ldots n-m+1}$, etc. The chance of a player who wishes to win at least with one of m cards $=$ the sum of all the above values of the series with m successively set equal to 1, 2, 3, etc., i.e. $= \frac{m}{n} - \frac{m.m-1}{1.2} \times \frac{1}{n.n-1} + \frac{m.m-1.m-2}{1.2.3} \times \frac{1}{n.n-1.n-2} - \frac{m.m-1.m-2.m-3}{1.2.3.4} \times \frac{1}{n.n-1.n-2.n-3} + \ldots$ up to $\pm \frac{m.m-1.m-2 \ldots m-m+1}{1.2.3 \ldots m} \times \frac{1}{n.n-1 \ldots n-m+1}$. Therefore, putting $m = n$, the chance of the player $= 1 - \frac{1}{1.2} + \frac{1}{1.2.3} - \frac{1}{1.2.3.4} + \ldots$ up to $\pm \frac{1}{1.2.3 \ldots n}$.

The DOCTRINE of CHANCES.   59

third Game, when it is eftimated before the Play begins,
is $\frac{aa}{a+b|^2} \times \frac{a-b}{a+b}$ &c.

Fourthly, Wherefore the Gain of the Hand of A is an infinite Series, *viz.* $1 + \frac{a}{a+b} + \frac{aa}{a+b|^2} + \frac{a^3}{a+b|^3} + \frac{a^4}{a+b|^4}$ &c.
to be Multiplyed by $\frac{a-b}{a+b}$. But the fum of that infinite
Series is $\frac{a+b}{b}$; Wherefore the Gain of the Hand of A
is $\frac{a+b}{b} \times \frac{a-b}{a+b} = \frac{a-b}{b}$.

Corollary I. If A has the advantage of the Odds, and B
Sets his Hand out, the Gain of A is the difference of the
numbers expreffing the Odds divided by the leffer. Thus
if A has the Odds of Five to Three, then his Gain will
be $\frac{5-3}{3} = \frac{2}{3}$.

Corollary II. If B has the Difadvantage of the Odds, and
A Sets his Hand out, the Lofs of B will be the difference
of the number expreffing the Odds divided by the greater :
Thus if B has but Three to Five of the Game, his Lofs will
be $\frac{2}{5}$.

Corollary III. If A and B do mutually engage to Set to
one-another as long as either of them wins without interruption, the Gain of A will be found to be $\frac{aa-bb}{ab}$: That
is the fum of the numbers expreffing the Odds Multiplyed
by their difference, the product of that Multiplication being
divided by the Product of the numbers expreffing the Odds.
Thus if the Odds were as Five to Three, the fum of 5 and
3 is 8, and the difference 2 ; Multiply 8 by 2, and the
Product 16 being divided by 15 (Product of the number
expreffing the Odds) the Quotient will be $\frac{16}{15}$, or $1\frac{1}{15}$,
which therefore will be the Gain of A.

PROBLEM XXV.

ANY *given number of Letters* a, b, c, d, e, f *&c. all of them*
different, being taken promifcuoufly, as it Happens : To find
the Probability that fome of them fhall be found in their places,
according.

60   *The* DOCTRINE *of* CHANCES.

according to the rank they obtain in the Alphabet ; *and that others of them shall at the same time be found out of their places.*

SOLUTION.

LET the number of all the Letters be $= n$; let the number of those that are to be in their places be $= p$, and the number of those that are to be out of their places $= q$. Suppose for Brevity sake $\frac{1}{n} = r$, $\frac{1}{n \times n-1} = s$, $\frac{1}{n \times n-1 \times n-2} = t$, $\frac{1}{n \times n-1 \times n-2 \times n-3} = v$ &c. then let all the Quantities 1, r, s, t, v &c. be written down with Signs alternately positive and negative, beginning at 1, if p be $= 0$; at r, if $p = 1$; at s, if $p = 2$ &c. Prefix to these Quantities the respective Coefficients of a Binomial Power, whose Index is $= q$: This being done, those Quantities taken all together will express the Probability required; thus the Probability that in Six Letters taken promiscuously, two of them, *viz. a* and *b* shall be in their places, and three of them, *viz. c, d, e* out of their places, will be

$$\frac{1}{6 \times 5} - \frac{3}{6 \times 5 \times 4} + \frac{3}{6 \times 5 \times 4 \times 3} - \frac{1}{6 \times 5 \times 4 \times 3 \times 2} = \frac{11}{720},$$

And the Probability that *a* shall be in its place, and *b, c, d, e* out of their places, will be

$$\frac{1}{6} - \frac{4}{6 \times 5} + \frac{6}{6 \times 5 \times 4} - \frac{4}{6 \times 5 \times 4 \times 3} + \frac{1}{6 \times 5 \times 4 \times 3 \times 2} = \frac{53}{720}.$$

The Probability that *a* shall be in its place, and *b, c, d, e, f* out of their places, will be

$$\frac{1}{6} - \frac{5}{6 \times 5} + \frac{10}{6 \times 5 \times 4} - \frac{10}{6 \times 5 \times 4 \times 3} + \frac{5}{6 \times 5 \times 4 \times 3 \times 2}$$
$$- \frac{1}{6 \times 5 \times 4 \times 3 \times 2 \times 1} = \frac{44}{720}, \text{ or } \frac{11}{180}.$$

The Probability that *a, b, c, d, e, f* shall be all displaced is,

$$1 - \frac{6}{6} + \frac{15}{6 \times 5} - \frac{20}{6 \times 5 \times 4} + \frac{15}{6 \times 5 \times 4 \times 3} - \frac{6}{6 \times 5 \times 4 \times 3 \times 2}$$
$$+ \frac{1}{6 \times 5 \times 4 \times 3 \times 2 \times 1}, \text{ or } 1 - 1 + \frac{1}{2} - \frac{1}{6} + \frac{1}{24} - \frac{1}{120}$$
$$+ \frac{1}{720} = \frac{265}{720} = \frac{53}{144}.$$

Hence

Hence it may be concluded that the Probability that one or more of them will be found in their places is $1 - \frac{1}{2} + \frac{1}{6} - \frac{1}{24} + \frac{1}{120} - \frac{1}{720} = \frac{91}{144}$; and that the Odds that one or more of them will be fo found are as 91 to 53.

N. B. So many Terms of this laſt Series are to be taken as there are Units in *n*.

DEMONSTRATION.

THE number of Chances for the Letter *a* to be in the firſt place contains the number of Chances, by which *a* being in the firſt place, *b* may be in the fecond, or out of it: This is an Axiom of common Senſe, of the fame degree of Evidence as that the Whole is equal to all its Parts.

From this it follows, that if from the number of Chances that there are for *a* to be in the firſt place, there be ſub-ſtracted the number of Chances that there are for *a* to be in the firſt place, and *b* at the fame time in the fecond, there will remain the number of Chances, by which *a* being in the firſt place, *b* may be excluded the fecond.

For the fame reaſon it follows, that if from the number of Chances that there are for *a* and *b* to be refpectively in the firſt and fecond places, there be ſubtracted the number of Chances by which *a*, *b* and *c* may be refpectively in the firſt, fecond and third places; there will remain the number of Chances by which *a* being in the firſt and *b* in the fecond, *c* may be excluded the third place : And fo of the reſt.

Let $+ a'$ denote the Probability that *a* ſhall be in the firſt place, and let $- a'$ denote the Probability of its being out of it. Likewiſe let the Probabilities that *b* ſhall be in the fecond place or out of it be refpectively expreſt by $+ b''$ and $- b''$.

Let the Probability that, *a* being in the firſt place, *b* ſhall be in the fecond, be expreſt by $a' + b''$: Likewiſe let the Probability that *a* being in the firſt place, *b* ſhall be excluded the fecond, be expreſt by $a' - b''$.

Generally. Let the Probability there is, that as many as are to be in their proper places, ſhall be fo, and at the fame time that as many others as are to be out of their proper places

R   ſhall

fhall be fo found, be denoted by the particular Probabilities of their being in their proper places, or out of them, written all together: So that for Inftance $a' + b'' + c''' - d'''' - e''''$ may denote the Probability that a, b and c fhall be in their proper places, and that at the fame time both d and e fhall be excluded their proper places.

Now to be able to derive a proper conclufion by vertue of this Notation, it is to be obferved, that of the Quantities which are here confidered, thofe from which the Subtraction is to be made, are indifferently compofed of any number of Terms connected by $+$ and $-$; the Quantities which are to be fubtracted do exceed by one Term thofe from which the fubtraction is to be made; the reft of the Terms being alike and their figns alike: And the remainder will contain all the Quantities that are alike with their own figns, and alfo the Quantity Exceeding, but with its fign varied.

It having been demonftrated in what we have faid of Permutations and Combinations, that $d' = \frac{1}{n}$, $a' + b'' = \frac{1}{n \times n - 1}$, $a' + b'' + c''' = \frac{1}{n \times n - 1 \times n - 2}$, let $\frac{1}{n}$, $\frac{1}{n \times n - 1}$ &c. be refpectively called r, s, t, v &c. This being fuppofed, we may come to the following conclufions.

$$b'' = r$$
$$b'' + a' = s$$

Therefore $\overline{b'' - a' = r - s}$

$$c''' + b'' = s \quad \text{for the fame reafon that } a' + b'' = s$$
$$c''' + b'' + a' = t$$

2° Theref. $\overline{c''' + b'' - a' = s - t}$

$$c''' - a' = r - s \qquad \text{By the firft Conclufion.}$$
$$c''' - a' + b'' = s - t \qquad \text{By the 2d.}$$

3° Theref. $\overline{c''' - a' - b'' = r - 2s + t}$

$$d'''' + c''' + b'' = t$$
$$d'''' + c''' + b'' + a' = v$$

4° Theref. $\overline{d'''' + c''' + b'' - a' = t - v}$

$$d'''' + c''' - a' = s - t \qquad \text{By the 2d. Conclufion.}$$
$$d'''' + c''' - a' + b'' = t - v \quad \text{By the 4th.}$$

5° Theref. $\overline{d'''' + c''' - a' - b'' = s - 2t + v}$

$$
\begin{array}{ll}
d'''' - b'' - a' = r - 2s + t & \text{By the } 3d. \text{ Conc.}\\
d'''' - b'' - a' + c''' = \quad\quad s - 2t + v & \text{By the } 5th.
\end{array}
$$

6° Theref. $d'''' - b'' - a' - c''' = r - 3s + 3t - v$

By the fame procefs, if no Letter be particularly affigned to be in its place, the Probability that fuch of them as are affigned may be out of their places will likewife be found thus.

$$
\begin{array}{ll}
- a' = 1 - r & \text{For} + a' \text{ and } - a' \text{ together make}\\
- a' + b'' = \quad r - s & \text{[Unity.}
\end{array}
$$

7° Theref. $- a' - b'' = 1 - 2r + s$

$$
\begin{array}{ll}
- a' - b'' = 1 - 2r + s & \text{By the } 7th. \text{ Conc.}\\
- a' - b'' + c'' = \quad\quad r - 2s + t & \text{By the } 3d. \text{ Conc.}
\end{array}
$$

8° Theref. $- a' - b'' - c'' = 1 - 3r - 3s - t$

Now examining carefully all the foregoing Conclufions, it will be perceived, that when the Queftion runs barely upon the difplacing any given number of Letters without requiring that any other fhould be in its place, but leaving it wholly indifferent, then the vulgar Algebraic Quantities which lie on the right hand of the Equations, begin conftantly with Unity: It will alfo be perceived, that when one fingle Letter is affigned to be in its place, then thofe Quantities begin with r; and that when two Letters are affigned to be in their places, they begin with s, and fo on. Moreover 'tis obvious, that thefe Quantities change their figns alternately, and that the Numerical Coefficients which are prefixt to them are thofe of a Binomial Power, whofe Index is equal to the number of Letters which are to be difplaced.

PROBLEM XXVI.

ANY given number of *different Letters* a, b, c, d, e, f *&c. being each of them repeated a certain number of times, and taken promifcuoufly as it Happens: To find the Probability that of fome of thofe Sorts, fome one Letter of each may be found in its proper place, and at the fame time that of fome other Sorts, no one Letter be found in its place.*

SOLU-

The Determination of the Accuracy of Observations

Comments on Gauss (1816)

1 Introduction

This influential paper has been reviewed by a good many writers, including Hald (1998, pp. 455–459) to whom we refer the reader. However, we must make a few points here.

Almost defensively Gauss begins by justifying his interest in the constant h occurring in the form (A) of the error law. He had put forward this form (Gauss, 1809) in the course of his famous "first proof" of the method of least squares and it remained in common use for the next century. Gauss had also pointed out that h could be considered as a measure of precision. In modern notation, h, later termed the *modulus of precision* or just the *precision*, is given by $h = 1/\sigma\sqrt{2}$, where σ is the standard deviation.

Without reference to Bessel, who had introduced the term *probable error* in 1815 (Hald, 1998, p. 360), Gauss uses this term, with symbol r defined by

$$\int_{-r}^{r} \frac{h}{\sqrt{\pi}} e^{-h^2\Delta^2} d\Delta = \frac{1}{2}.$$

Thus $r = \Phi^{-1}(0.75)\sigma$, where Φ denotes the standard normal cdf. Also needed is the constant $\rho = rh = \Phi^{-1}(0.75)/\sqrt{2} = 0.4769363$.

2 Estimation of h and r by Inverse Probability

We write $x_1, x_2, \ldots$ for Gauss's $\alpha, \beta, \ldots$ in Section 3, and write $\mathbf{x}$ for $(x_1, \ldots, x_m)$. Gauss refers to his *Theoria motus corporum coelestium* (1809) which was translated into English by the Commander of the U.S. Navy (Davis, 1857). In the cited Section 176 Gauss uses an ad hoc Bayesian-type argument to show that the posterior density of h is proportional to the likelihood (in modern terminology):

$$f(h|\mathbf{x}) \propto f(\mathbf{x}|h), \tag{a}$$

where he has assumed h to be uniformly distributed over $[0, \infty)$. Gauss mentions neither Bayes nor Laplace, although the latter had popularized this approach since Laplace (1774).

Maximizing $f(h|\mathbf{x})$ w.r.t. h Gauss obtains here what Edwards (1974) has called the maximum probability estimator

$$\hat{h} = (m/2\Sigma x_i^2)^{1/2}.$$

Of course, $\hat{h}$ is also the maximum likelihood estimate, but the distinction between the methods of maximum probability and maximum likelihood is conceptually important. Since $r = \rho/h$, Gauss states it follows that $\hat{r} = \rho/\hat{h}$. He notes that these results hold for any sample size, but from here on takes m to be large.

By an ingenious argument Gauss goes on to show that, for large m, the true value of h lies with probability $\frac{1}{2}$ in

$$\left(\hat{h}(1 - \rho m^{-1/2}), \ \hat{h}(1 + \rho m^{-1/2})\right). \tag{b}$$

This is his eq. (B). (Here and elsewhere Gauss's use of H for $\hat{h}$ is confusing.) In modern terminology (b) is a large-sample 50% Bayesian confidence interval corresponding to a uniform prior, and hence is also an ordinary large-sample CI. Similar remarks apply to (C) on replacing R by $\hat{r}$.

3   Use of Direct Probability

So far Gauss has employed what we now call Bayesian methods, long known as inverse probability methods. From given data he has made inferences about the distribution of h by use of (a). Now he turns to the direct probability counterpart, i.e., sampling theory: given the pdf $\phi(x)$, including an unknown parameter (h or r), to find the distribution and properties of various estimators of the parameter of interest. This parameter he takes to be r, the constant multiple of the standard deviation. Hald (1998, p. 456) writes that Gauss henceforth, in all his work, exclusively uses sampling theory.

Gauss confines himself almost entirely to the case of large m, making repeated use of the new central limit theorem. Like his contemporaries sparing in giving credit to others, Gauss does acknowledge Laplace here. Finding asymptotic 50% confidence intervals for r, based on $S_n = \Sigma x_i^n$, he shows numerically that the length of the CI drops from $n = 1$ to $n = 2$ and then increases for $n = 3, 4, 5, 6$.

Although 114 observations are required with S_1 to 100 for S_2, Gauss points out that use of S_1 is more convenient. In fact, ease of calculation is so important that Gauss even considers $M = \text{med}|x_i|$, the median absolute deviation (MAD). Without proof he is able to give the correct expression for the asymptotic 50% CI of r based on M (eq. (D)), but makes a rare numerical error in evaluating it. Hauber (1830) provides a rough proof and Encke (1832) a more careful one, for which he thanks Dirichlet. In both papers, Gauss's figure is quietly corrected without pointing out his error! It turns out that 272 observations are needed with M (when 100 are needed with S_2) rather than Gauss's 249. See Harter's (1978) entries on the three papers.

That S_1 and especially M have the advantage of greater robustness than $\sqrt{S_2}$ does not enter at this early stage.

4   Notes on this Translation

A previous English translation of this paper is part of a technical report from Princeton University, by Trotter (1957). This report is a translation from the French of Bertrand (1855), itself a translation, authorized by Gauss, of Gauss's work on the method of least squares. Gauss originally published mainly in Latin and partly in German. Although both translators have performed a valuable service, it seemed desirable to prepare a new translation directly from the German.

This has also enabled us to correct a mistake in translation made by Bertrand at the end of Section 7 and preserved by Trotter. Referring to the use of M, Gauss writes: "This procedure is therefore only little more accurate than the use of formula 6." Bertrand retains Gauss's erroneous 249 figure, but translates his "wenig genauer" ("little more accurate") by "pas beaucoup moins exact" ("not much less exact"). Curiously, the incorrect translation becomes true, once Gauss's error is taken into account.

To reduce confusion we have, following Trotter, replaced Gauss's Θt in his Section 2 by $\Theta(t)$, etc., and in Section 5 his ϕx by $\phi(x)$, and his K^n, S^n by K_n, S_n.

References

Bertrand, J. (1855). *Méthode des moindres carrés. Mémoire sur la combinaison des observations, par Ch.-Fr. Gauss.* Mallet-Bachelier, Paris.

Davis, C.H. (1857). *Theory of the Motion of Heavenly Bodies.* Translation of Gauss (1809). Little, Brown, Boston.

Edwards, A.W.F. (1974). The history of likelihood. *Internat. Statist. Rev.,* **42**, 9–15.

Encke, J. F. (1832). Über die Methode der kleinsten Quadrate. *Berliner astronomisches Jahrbuch für* 1834.

Gauss, C.F. (1809). *Theoria motus corporum coelestium.* In *Werke,* **7**, 1–280. Translated by Davis (1857).

Gauss, C.F. (1816). Bestimmung der Genauigkeit der Beobachtungen. In *Werke,* **4**, 109–117.

Gauss, C.F. (1863–1933). *Werke*, 12 vols., Königliche Gesellschaft der Wissenschaften zu Göttingen.

Hald, A. (1998). *A History of Mathematical Statistics from 1750 to 1930.* Wiley, New York.

Harter, H.L. (1978). *A Chronological Annotated Bibliography of Order Statistics*, Vol. 1: Pre-1950. U.S. Government Printing Office. [Reprinted by American Sciences Press, Syracuse, New York.]

Hauber, C.F. (1830). Über die Bestimmung der Genauigkeit der Beobachtungen. *Zeit. Physik u. Mathematik*, **7**, 286–314.

Laplace, P. S. (1774). Mémoire sur la probabilité des causes par les événments. *Mém. Acad. R. Sci.*, **6**, 621–656.

Trotter, H.F. (1957). *Gauss's Work (1803–1826) on the Theory of Least Squares.* English translation of Bertrand (1855). Tech. Rep. No. 5. Statistical Techniques Research Group, Princeton University.

The Determination of the Accuracy of Observations

C. F. Gauss

1

When making the case for the so-called method of least squares, it is assumed that the probability of an error of observation Δ may be expressed by the formula

$$\frac{h}{\sqrt{\pi}} \cdot e^{-hh\Delta\Delta}, \tag{A}$$

where π is the semi-perimeter of the unit circle, e is the base of natural logarithms, and h is also a constant that according to Section 178 of *Theoria Motus Corporum Coelestium* may be regarded as a measure of the accuracy of the observations. It is not at all necessary to know the value of h in order to apply the method of least squares to determine the most probable values of those quantities [parameters] on which the observations depend. Also, the ratio of the accuracy of the results to the accuracy of the observations does not depend on h. However, knowledge of its value is in itself interesting and instructive, and I will therefore show how we can arrive at such knowledge through the observations themselves.

2

I begin with a few explanatory remarks on the subject. For the sake of brevity I denote by $\Theta(t)$ the value of the integral

$$\int \frac{2e^{-tt}}{\sqrt{\pi}} dt,$$

calculated from $t = 0$. A few individual values will give an indication of the behavior of this function. We have

$$
\begin{array}{rcccl}
0.5000000 & = & \Theta(0.4769363) & = & \Theta(\rho) \\
0.6000000 & = & \Theta(0.5951161) & = & \Theta(1.247790\rho) \\
0.7000000 & = & \Theta(0.7328691) & = & \Theta(1.536618\rho) \\
0.8000000 & = & \Theta(0.9061939) & = & \Theta(1.900032\rho) \\
0.8427008 & = & \Theta(1) & = & \Theta(2.096716\rho) \\
0.9000000 & = & \Theta(1.1630872) & = & \Theta(2.438664\rho) \\
0.9900000 & = & \Theta(1.8213864) & = & \Theta(3.818930\rho) \\
0.9990000 & = & \Theta(2.3276754) & = & \Theta(4.880475\rho) \\
0.9999000 & = & \Theta(2.7510654) & = & \Theta(5.768204\rho) \\
1 & = & \Theta(\infty). & &
\end{array}
$$

The probability that the error of an observation lies between the limits $-\Delta$ and Δ, or that, disregarding the sign, it does not exceed Δ, is

$$
\int \frac{h e^{-hhxx}\, dx}{\sqrt{\pi}}
$$

if we extend the integral from $x = -\Delta$ to $x = +\Delta$; or the probability equals twice the same integral taken from $x = 0$ to $x = \Delta$, and consequently $= \Theta(h\Delta)$.

The probability that the error is not less than ρ/h is therefore $= 1/2$, in other words, equal to the probability of the contrary event. We will call this value, ρ/h, the *probable error* and will denote it by r. On the other hand, the probability that the error exceeds $2.438664r$ is only $1/10$, the probability that the error increases beyond $3.818930r$ is only $1/100$, and so on.

3

Let us now assume that for m actual observations the errors $\alpha, \beta, \gamma, \delta, \ldots$ were committed, and let us investigate what may be inferred from this regarding the value of h and r. If we make two assumptions by taking the true value of h either $= H$ or $= H'$, then the probabilities with which the errors $\alpha, \beta, \gamma, \delta, \ldots$ may be expected, are respectively as

$$
H e^{-HH\alpha\alpha} . H e^{-HH\beta\beta} . H e^{-HH\gamma\gamma} \ldots
$$

is to

$$
H' e^{-H'H'\alpha\alpha} . H' e^{-H'H'\beta\beta} . H' e^{-H'H'\gamma\gamma} \ldots
$$

i.e., as

$$
H^m e^{-HH(\alpha\alpha+\beta\beta+\gamma\gamma+\cdots)} \quad \text{is to} \quad H'^m e^{-H'H'(\alpha\alpha+\beta\beta+\gamma\gamma+\cdots)}.
$$

Consequently, the same relation holds for the probabilities that H or H' was the true value of h *after* the occurrence of the above errors (T.M.C.C., Section 176). In other words, the probability of any value of h is proportional to the quantity

$$h^m c^{-hh(\alpha\alpha+\beta\beta+\gamma\gamma+\cdots)}.$$

The *most probable* value of h is consequently that value for which this quantity is maximized, which according to well-known rules

$$= \sqrt{\frac{m}{2(\alpha\alpha + \beta\beta + \gamma\gamma + \cdots)}}.$$

The most probable value of r is consequently

$$= \rho\sqrt{\frac{2(\alpha\alpha + \beta\beta + \gamma\gamma + \cdots)}{m}}$$

$$= 0.6744897\sqrt{\frac{(\alpha\alpha + \beta\beta + \gamma\gamma + \cdots)}{m}}.$$

This result holds generally, whether m is large or small.

4

It is easy to comprehend that in this determination of h and r we are the less entitled to expect much accuracy, the smaller m is. Let us, therefore, develop an expression for the degree of precision we can attach to this determination in the case when m is a large number. We denote the previously found [most] probable error of h, namely

$$\sqrt{\frac{m}{2(\alpha\alpha + \beta\beta + \gamma\gamma + \cdots)}},$$

by H, for the sake of brevity, and note that the probability that H is the true value of h stands to the probability that the true value $= H + \lambda$ in the ratio

$$H^m e^{-\frac{m}{2}} : (H + \lambda)^m e^{-\frac{m(H+\lambda)^2}{2HH}}$$

or as

$$1 : e^{-\frac{\lambda\lambda m}{HH}\left(1 - \frac{1}{3}\cdot\frac{\lambda}{H} + \frac{1}{4}\cdot\frac{\lambda\lambda}{HH} - \frac{1}{5}\cdot\frac{\lambda^3}{H^3}\cdots\right)}.$$

The second member is significant compared to the first only if λ/H is a small fraction. Hence, we may permit ourselves to use the ratio

$$1 : e^{-\frac{\lambda\lambda m}{HH}}$$

in place of the above. Now this really means the following: the probability that the true value of h lies between $H + \lambda$ and $H + \lambda + d\lambda$ is very nearly

$$= Ke^{-\frac{\lambda\lambda m}{HH}}\, d\lambda,$$

where K is a constant that must be determined so that the integral

$$\int Ke^{-\frac{\lambda\lambda m}{HH}}\, d\lambda,$$

taken between the permissible limits of λ, becomes 1. In place of such limits, it is permissible here to take these as $-\infty$ and $+\infty$ since, in view of the size of m,

$$e^{-\frac{\lambda\lambda m}{HH}}$$

evidently becomes insignificant as soon as λ/H is no longer a small fraction. Then

$$K = \frac{1}{H}\sqrt{\frac{m}{\pi}},$$

and, therefore, the probability that the true value of h lies between $H - \lambda$ and $H + \lambda$ becomes

$$\Theta\left(\frac{\lambda}{H}\sqrt{m}\right).$$

So, this probability $= 1/2$ if

$$\frac{\lambda}{H}\sqrt{m} = \rho.$$

It is therefore an even bet that the true value of h lies between

$$H\left(1 - \frac{\rho}{\sqrt{m}}\right) \quad \text{and} \quad H\left(1 + \frac{\rho}{\sqrt{m}}\right) \tag{B}$$

or that the true value of r falls between

$$\frac{R}{1 - \frac{\rho}{\sqrt{m}}} \quad \text{and} \quad \frac{R}{1 + \frac{\rho}{\sqrt{m}}},$$

if we denote by R the most probable value of r obtained in the previous section. We may call these limits the *probable limits of the true values of h and r*. For the probable limits of the true value of r we may evidently here also put

$$R\left(1 - \frac{\rho}{\sqrt{m}}\right) \quad \text{and} \quad R\left(1 + \frac{\rho}{\sqrt{m}}\right). \tag{C}$$

5

In the preceding investigation we started from the point of view that we regarded $\alpha, \beta, \gamma, \delta, \ldots$ as definite and given values and that we were seeking the value of the probability that the true error of h or r lies between certain limits. The matter may also be viewed from a different angle: under the supposition that the errors of observation are subject to some definite law of probability, we can determine the probability with which the sum of the squares of m errors of observation falls between certain limits. This problem has already been solved by Laplace under the condition that m is large, as has the problem of seeking the probability that the sum itself of m observations falls between certain limits. This investigation can easily be further generalized. I content myself here with indicating the result.

Let $\phi(x)$ denote the probability of the observational error x, so that $\int \phi(x) \cdot dx$ becomes 1 if the integral is extended from $x = -\infty$ to $x = +\infty$. For these same limits let us generally denote the value of the integral

$$\int \phi(x) \cdot x^n dx$$

by K_n. Further, let S_n be the sum

$$\alpha^n + \beta^n + \gamma^n + \delta^n + \cdots,$$

where $\alpha, \beta, \gamma, \delta, \ldots$ denote m arbitrary errors of observation. The terms of the above sum are all to be taken as positive, also for n odd.

Then mK_n is the most probable value of S_n and the probability that the true value of S_n falls between the limits $mK_n - \lambda$ and $mK_n + \lambda$

$$= \Theta \left(\frac{\lambda}{\sqrt{[2m(K_{2n} - K_nK_n)]}} \right).$$

Consequently, the probable limits of S_n are

$$mK_n - \rho\sqrt{[2m(K_{2n} - K_nK_n)]} \quad \text{and} \quad mK_n + \rho\sqrt{[2m(K_{2n} - K_nK_n)]}.$$

This result is valid for any law of error of observations. If we apply it to the case

$$\phi(x) = \frac{h}{\sqrt{\pi}} e^{-hhxx},$$

we find

$$K_n = \frac{\Pi(\frac{1}{2}(n-1))}{h^n \sqrt{\pi}};$$

the symbol Π is taken with its meaning in the *Disquisitiones generales circa seriem infinitum* (Comm. Nov. Soc. Gotting, Vol. II, Note 5, Section 28).

[Translator: $\Pi(\frac{1}{2}(n-1)) = \Gamma(\frac{1}{2}(n+1))$.] Hence,

$$K = 1, \quad K_1 = \frac{1}{h\sqrt{\pi}}, \quad K_2 = \frac{1}{2hh}, \quad K_3 = \frac{1}{h^3\sqrt{\pi}},$$

$$K_4 = \frac{1.3}{4h^4}, \quad K_5 = \frac{1.2}{h^5\sqrt{\pi}}, \quad K_6 = \frac{1.3.5}{8h^6}, \quad K_7 = \frac{1.2.3}{h^7\sqrt{\pi}}, \quad \text{etc.}$$

It follows that the most probable value of S_n is

$$\frac{m\Pi(\frac{1}{2}(n-1))}{h^n\sqrt{\pi}}$$

and that the most probable limits of the true value of S_n are

$$\frac{m\Pi(\frac{1}{2}(n-1))}{h^n\sqrt{\pi}} \left\{ 1 - \rho\sqrt{\left(\frac{2}{m} \cdot \left(\frac{\Pi(n-\frac{1}{2}) \cdot \sqrt{\pi}}{(\Pi\frac{1}{2}(n-1))^2} - 1 \right) \right)} \right\}$$

and

$$\frac{m\Pi(\frac{1}{2}(n-1))}{h^n\sqrt{\pi}} \left\{ 1 + \rho\sqrt{\left(\frac{2}{m} \cdot \left(\frac{\Pi(n-\frac{1}{2}) \cdot \sqrt{\pi}}{(\Pi\frac{1}{2}(n-1))^2} - 1 \right) \right)} \right\}.$$

If, as above, we set $\rho/h = r$, so that r represents the probable error of observation, then the most probable value of

$$\rho\sqrt[n]{\frac{S_n\sqrt{\pi}}{m\Pi(\frac{1}{2}(n-1))}}$$

is clearly r. Also, the probable limits of the value of r are

$$r \left\{ 1 - \frac{\rho}{n}\sqrt{\left(\frac{2}{m} \cdot \left(\frac{\Pi(n-\frac{1}{2}) \cdot \sqrt{\pi}}{(\Pi\frac{1}{2}(n-1))^2} - 1 \right) \right)} \right\}$$

and

$$r \left\{ 1 + \frac{\rho}{n}\sqrt{\left(\frac{2}{m} \cdot \left(\frac{\Pi(n-\frac{1}{2}) \cdot \sqrt{\pi}}{(\Pi\frac{1}{2}(n-1))^2} - 1 \right) \right)} \right\}.$$

Hence, it is also an even bet that r lies between the limits

$$\rho\sqrt[n]{\frac{S_n\sqrt{\pi}}{m\Pi\frac{1}{2}(n-1)}} \left\{ 1 - \frac{\rho}{n}\sqrt{\left(\frac{2}{m} \cdot \left(\frac{\Pi(n-\frac{1}{2}) \cdot \sqrt{\pi}}{(\Pi\frac{1}{2}(n-1))^2} - 1 \right) \right)} \right\}$$

and

$$\rho\sqrt[n]{\frac{S_n\sqrt{\pi}}{m\Pi\frac{1}{2}(n-1)}} \left\{ 1 + \frac{\rho}{n}\sqrt{\left(\frac{2}{m} \cdot \left(\frac{\Pi(n-\frac{1}{2}) \cdot \sqrt{\pi}}{(\Pi\frac{1}{2}(n-1))^2} - 1 \right) \right)} \right\}.$$

For $n = 2$ these limits are

$$\rho\sqrt{\frac{2S_2}{m}} \left\{ 1 - \frac{\rho}{\sqrt{m}} \right\} \quad \text{and} \quad \rho\sqrt{\frac{2S_2}{m}} \left\{ 1 + \frac{\rho}{\sqrt{m}} \right\},$$

in complete agreement with those found above (Section 4). In general, we have the limits

$$\rho\sqrt{2}\cdot\sqrt[n]{\frac{S_n}{m.1.3.5.7\ldots(n-1)}}$$
$$\left\{1-\frac{\rho}{n}\sqrt{\left(\frac{2}{m}\cdot\left(\frac{(n+1).(n+3)\ldots(2n-1)}{1.3.5\ldots(n-1)}-1\right)\right)}\right\}$$

$$\rho\sqrt{2}\cdot\sqrt[n]{\frac{S_n}{m.1.3.5.7\ldots(n-1)}}$$
$$\left\{1+\frac{\rho}{n}\sqrt{\left(\frac{2}{m}\cdot\left(\frac{(n+1).(n+3)\ldots(2n-1)}{1.3.5\ldots(n-1)}-1\right)\right)}\right\}$$

for n even, and

$$\rho\sqrt[n]{\frac{S_n\sqrt{\pi}}{m.1.2.3\ldots\frac{1}{2}(n-1)}}\left\{1-\frac{\rho}{n}\sqrt{\left(\frac{1}{m}\cdot\left(\frac{1.3.5.7\ldots(2n-1)\pi}{(2.4.6\ldots(n-1))^2}-2\right)\right)}\right\}$$

$$\rho\sqrt[n]{\frac{S_n\sqrt{\pi}}{m.1.2.3\ldots\frac{1}{2}(n-1)}}\left\{1+\frac{\rho}{n}\sqrt{\left(\frac{1}{m}\cdot\left(\frac{1.3.5.7\ldots(2n-1)\pi}{(2.4.6\ldots(n-1))^2}-2\right)\right)}\right\}$$

for n odd.

6

I will still add the numerical values for the simplest cases:

Probable Limits of r

I. $0.8453473\ \dfrac{S_1}{m}\cdot\left(1+\dfrac{0.5095841}{\sqrt{m}}\right)$

II. $0.6744897\ \sqrt{\dfrac{S_2}{m}}\cdot\left(1\mp\dfrac{0.4769369}{\sqrt{m}}\right)$

III. $0.5771897\ \sqrt[3]{\dfrac{S_3}{m}}\cdot\left(1\mp\dfrac{0.4971987}{\sqrt{m}}\right)$

IV. $0.5125017\ \sqrt[4]{\dfrac{S_4}{m}}\cdot\left(1\mp\dfrac{0.5507186}{\sqrt{m}}\right)$

V. $0.4655532\ \sqrt[5]{\dfrac{S_5}{m}}\cdot\left(1\mp\dfrac{0.6355080}{\sqrt{m}}\right)$

VI. $0.4294972\ \sqrt[6]{\dfrac{S_6}{m}}\cdot\left(1+\dfrac{0.7557764}{\sqrt{m}}\right).$

So we see from this also that the second mode of determination is the most advantageous of all. Indeed, a hundred errors of observations, treated according to this formula, provide an equally reliable result as

114 acc. to I, 109 acc. to III, 133 acc. to IV, 178 acc. to V, 251 acc. to VI.

On the other hand, formula I has the advantage of being the most convenient of all to calculate. Since this formula is only a little less precise than II, one may wish to employ it if one does not already otherwise know the sum of squares of the errors, or does not wish to know it.

7

Even more convenient, although considerably less accurate, is the following procedure: order all m errors of observation according to their absolute values, and denote by M the middle one if their number is odd, or the arithmetic mean of the two middle ones for an even number. It may be shown, but we cannot elaborate here, that for a large number of observations the most probable value of M is r, and that the probable limits of M are

$$r\left(1 - e^{\rho\rho}\sqrt{\frac{\pi}{8m}}\right) \quad \text{and} \quad r\left(1 + e^{\rho\rho}\sqrt{\frac{\pi}{8m}}\right);$$

or the probable limits of r are

$$M\left(1 - e^{\rho\rho}\sqrt{\frac{\pi}{8m}}\right) \quad \text{and} \quad M\left(1 + e^{\rho\rho}\sqrt{\frac{\pi}{8m}}\right), \tag{D}$$

or numerically

$$M\left(1 \mp \frac{0.7520974}{\sqrt{m}}\right).$$

This procedure is, therefore, only a little more accurate than the use of formula 6. Also, we would need to consult 249 errors of observation in order to get as far as with 100 observations for formula 2.

8

The application of some of these methods to the errors of 48 observations by Bessel of the right ascension of the polar star, reported on p. 234 of Bode's astronomical yearbook for 1818, gave

$$S_1 = 60.46'', \quad S_2 = 110.600'', \quad S_3 = 250.341118''.$$

From this the probable values of r followed:

acc. to formula	I	1.065″	probable error =	±0.078″
	II	1.024″		±0.070″
	III	1.001″		±0.072″
acc. to Sect. 7		1.045″		±0.113″.

Such a degree of agreement could hardly be expected. Bessel himself gives 1.067″ and, therefore, seems to have calculated according to formula I.

The Introduction of Asymptotic Relative Efficiency

Comments on Laplace (1818)

1 Introduction

In this extract from the second supplement of his famous *Théorie analytique des probabilités*, Laplace makes, in the context of simple linear regression, a large-sample comparison of what we now call L_1- and L_2- estimation. He essentially introduces the notion of asymptotic relative efficiency and, incidentally, pioneers the theory of order statistics.

Earlier, Laplace (1810, 1812) had obtained the all-important central limit theorem without, however, recognizing that it does not hold for certain parent populations. Laplace now uses the theorem to show that both the least-squares estimator of the regression coefficient and a weighted median estimator are asymptotically normally distributed. He then compares the two asymptotic variances. Laplace even obtains the joint asymptotic distribution of the estimators and shows that if the underlying distribution is normal, then no linear combination of the two estimators can improve on the sole use of the least-squares estimator. A special case of this result is that no linear combination of the mean and median of a normal sample can improve on the mean as an estimator of the population mean. Thus Laplace partially anticipates results obtained in Fisher (1920).

To understand the extract translated here several preliminaries are needed. The "preceding methods" of the opening sentence refer to the simple linear model (of regression through the origin) that may be written as the system (A). Here $\epsilon_1, \ldots, \epsilon_n$ are independent error terms with "probability law" $\phi(\epsilon)$. However, $\phi(\epsilon)$ needs to be divided by $2k = \int_{-\infty}^{\infty} \phi(\epsilon)d\epsilon$ to be converted to a density function. The system (A) constitutes the *equations of condition*, a term sometimes applied also to the (generally) inconsistent equations obtained by setting $\epsilon_i = 0$, $i = 1, \ldots, n$. Multiplying the ith equation of condition by x_i and summing, we have $\beta = (\Sigma x_i y_i + \Sigma x_i \epsilon_i)/\Sigma x_i^2$. Setting the ϵ_i equal to zero gives the least-squares estimate $\hat{\beta} = \Sigma x_i y_i / \Sigma x_i^2$.

Laplace calls this procedure *the most advantageous method* and goes on to compare it with *the method of situation* which he considers in detail.

Since Laplace's notation is very confusing to the modern reader, we have replaced his a_i, p_i, x_i, y by y_i, x_i, ϵ_i, β. Following Laplace (and all 19th century authors), we make no distinction here between random variables and their realizations.

2 The Method of Situation, the Weighted Median, and Order Statistics

Preceding Legendre's 1805 publication of the method of least squares by almost 50 years, the Serbian scientist Roger Boscovich (1711–1787) put forward a method of fitting a straight line to the points $(x_1, y_1), \ldots, (x_n, y_n)$ that minimizes the sum of the absolute (vertical) deviations $\sum_{i=1}^{n} |\epsilon_i|$ subject to $\Sigma \epsilon_i = 0$ (see Eisenhart, 1961; Sheynin, 1973; Stigler, 1973; Hald, 1998, pp. 112–116). Laplace drops the condition and provides a clear algebraic proof in place of Boscovich's geometric argument. Actually, Laplace is repeating here a proof he has already presented twice before (Laplace, 1793, 1799). Only the first time does he mention Boscovich.

As Laplace points out, for the special case $x_1 = \cdots = x_n = 1$, n odd, the estimate y_r/x_r reduces to the median, by virtue of (B). Thus in the general case y_r/x_r may be regarded as a weighted median. Although only regression through the origin is treated here, the minimization argument goes through unchanged upon replacing x_i by $x_i - \bar{x}$, and y_i by $y_i - \bar{y}$, $i = 1, \ldots, n$.

Now come a number of important new developments. Laplace proceeds to obtain the exact distribution of y_r/x_r and seems first to use an argument now basic in the theory of order statistics. Although the special case $x_i = 1$, $i = 1, \ldots, n$, gives only the median (for n odd), the reasoning clearly applies to any order statistic. Laplace's language shows that he is pleased with his approach.

Of course, Laplace can not do much with his finite-sample result and hence, typically, turns to asymptotic theory. It is only at this stage that he uses the assumption $\phi(-\epsilon) = \phi(\epsilon)$. Specifically, he finds the asymptotic distribution of $\zeta = (y_r/x_r) - \beta$, arriving at expression (C), where he uses c for the base of natural logarithms. For the special case of the median, (C) reduces to the standard result, since $\phi(\epsilon)$ has to be divided by $2k$ to become a pdf.

3 Asymptotic Relative Efficiency

Laplace had noted previously that the difference $\hat{\beta} - \beta$ has an asymptotic distribution proportional to (D). (As pointed out above, this does not hold universally.) Denoting this difference also by ζ, he proceeds to compare the coefficients of ζ^2 in (C) and (D), thus coming very close to the notion of asymptotic relative efficiency.

Laplace's condition (E) for the weighted median to be preferable to the least-squares estimator is in modern notation just

$$f(0) > 1/(2\sigma),$$

where the parent density $f(x)$ is symmetric about zero, with variance σ^2. While not satisfied in the normal case, as shown by Laplace, the inequality does hold for the uniform distribution, for example. This implies in particular that in the latter case the sample median (although of course not best) is preferable to the sample mean as a large-sample estimator of the population mean. Laplace does not give any example other than the normal, but his language indicates that he regards (E) as possible. However, this does not deter him from continuing to call the least-squares method the most advantageous method.

4   The Joint Distribution of the Least-Squares and the Weighted Median Estimators

To conclude his Second Supplement, Laplace derives the joint asymptotic distribution of his two ζ's, now denoting the least-squares estimator by ζ'. The aim of this remarkable undertaking is to investigate conditions under which use of ζ', the most advantageous method, can be improved by use of a combination of ζ' and the method of situation ζ.

Formula (F) requires explanation, since it is hard to extricate from Laplace's reference to the second book of the *Analytical Theory*. Apart from a proportionality factor, the formula gives the exact joint density of $\zeta = \epsilon_r/x_r$ and $\ell = \zeta'\Sigma x_i^2 = \Sigma x_i\epsilon_i$, where r is determined by the condition (C).

The joint density of ζ and ℓ is

$$f(\zeta)f(\ell|\zeta) ,$$

where $f(\ell|\zeta) = \frac{1}{2\pi}\int_{-\infty}^{\infty} e^{-\ell\omega\sqrt{-1}}\psi(\ell|\zeta)d\omega$, $\psi(\ell|\zeta)$ being the characteristic function of $\ell|\zeta$. We have $\psi(\ell|\zeta) = E(e^{x_1\epsilon_1\omega\sqrt{-1}}\ldots e^{x_n\epsilon_n\omega\sqrt{-1}}|\zeta)$. Given ζ, conditions (C) mean that $\epsilon_i > \zeta x_i$, $i = 1,\ldots,r-1$ and $\epsilon_i < \zeta x_i$, $i = r+1,\ldots,n$. Recall that $\int_{-\infty}^{\infty} d\epsilon.\phi(\epsilon) = 2k$ and denote $\int_{-\infty}^{x} d\epsilon.\phi(\epsilon)$ by $\Phi(x)$. Conditionally, the ϵ_i are independent truncated rvs, so that

$$\psi(\ell|\zeta) \propto \left(\prod_{i=1}^{r-1}\int_{\zeta x_i}^{\infty} d\epsilon_i.\frac{\phi(\epsilon_i)}{2k - \Phi(\zeta x_i)}.e^{x_i\epsilon_i\omega\sqrt{-1}}\right)$$

$$\cdot e^{x_r^2 \zeta \omega \sqrt{-1}} \left(\prod_{i=r+1}^{n} \int_{-\infty}^{\zeta x_i} d\epsilon_i \cdot \frac{\phi(\epsilon_i)}{\Phi(\zeta x_i)} \cdot e^{x_i \epsilon_i \omega \sqrt{-1}} \right).$$

Again, from conditions (C),

$$f(\zeta) \propto \prod_{i=1}^{r-1} [2k - \Phi(\zeta x_i)] \phi(\zeta x_r) \prod_{i=r+1}^{n} \Phi(\zeta x_i).$$

Since the subscripts of the ϵ_i may be dropped, formula (F) follows, except that we show $e^{x_r^2 \zeta \omega \sqrt{-1}}$ where Laplace, by an apparent oversight, has $e^{x_r \zeta \omega \sqrt{-1}}$. This discrepancy is unimportant in Laplace's subsequent asymptotic analysis which is essentially straightforward but requires great care. See also Hald (1998, pp. 447–452) who, uncharacteristically however, is unsatisfactory on expression (F).

References

Eisenhart, C. (1961). Boscovich and the combination of observations. In: *Roger Joseph Boscovich*, L.L. Whyte, ed. Allen and Unwin, London, pp. 200–212. [Reprinted in Kendall and Plackett (1977), pp. 88–100.]

Fisher, R.A. (1920). A mathematical examination of the methods of determining the accuracy of an observation by the mean error, and by the mean square error. *Mon. Nat. R. Astron. Soc.*, **80**, 758–770. [Reprinted as Paper 2 in Fisher (1950).]

Fisher, R. A. (1950). *Contributions to Mathematical Statistics*. Wiley, New York.

Hald, A. (1998). *A History of Mathematical Statistics from 1750 to 1930*. Wiley, New York.

Kendall, M. G. and Plackett, R. L. (eds.) (1977). *Studies in the History of Statistics and Probability*. Vol. 2. Griffin, London.

Laplace, P.S. (1793). Sur quelques points du système du monde. *Mém. Acad. R. Sci. Paris (1789)*, 1–87. [Reprinted in *Oeuvres de Laplace*, **11**. Imprimerie Royale, Paris (1847).]

Laplace, P.S. (1799). *Traité de mécanique céleste*, Vol. 2, 3rd Book, Sections 40–41. [Translated into English by N. Bowditch in 1832 and reprinted in Laplace (1966, pp. 434–442, 448–450).]

Laplace, P.S. (1810). Mémoire sur les approximations des formules qui sont fonctions de trés grands nombres et sur leur application aux probabilités. *Mém. Acad. R. Sci. Paris*, 353–415. [Reprinted in *Oeuvres de Laplace*, **12**. Imprimerie Royale, Paris (1847).]

Laplace, P.S. (1812). *Théorie analytique des probabilités*. Courier, Paris. [Reprinted in *Oeuvres de Laplace*, **7**. Imprimerie Royale, Paris (1847).]

Laplace, P.S. (1818). *Théorie analytique des probabilités, deuxième sup--plément*. [Reprinted in *Oeuvres de Laplace*, **7**. Imprimerie Royale, Paris (1847).]

Laplace, P. S. (1966). Celestial Mechanics, Vol. 2, translated, with comme-naty, by N. Bodwitch. Chelsea, New York.

Legendre, A.M. (1805). *Nouvelles méthodes pour la détermination des orbites des comètes*. Courier, Paris. [Reprinted by Dover, New York, 1959.]

Sheynin, O.B. (1973). R.J. Boscovich's work on probability. *Arch. Hist. Exact Sci.*, **9**, 306–324.

Stigler, S.M. (1973). Laplace, Fisher, and the discovery of the concept of sufficiency. *Biometrika*, **60**, 439–445. [Reprinted in Kendall and Plackett (1977), pp. 271–277.]

On the Probability of Results Deduced by Methods of any Kind from a Large Number of Observations

P. S. Laplace

Section 2

The preceding methods reduce to multiplication of each equation of condition by a factor, and addition of all the products in order to form a final equation. But we can employ other considerations to obtain the desired result. For example, we can choose that equation of condition approaching most closely to the truth. The procedure I have given in Section 40 of *Mécanique céleste* is of this kind.

Suppose the equations of the previous section, namely

$$
\begin{aligned}
y_1 - \beta x_1 - \epsilon_1 &= 0 \\
y_2 - \beta x_2 - \epsilon_2 &= 0 \\
&\;\;\vdots \\
y_n - \beta x_n - \epsilon_n &= 0,
\end{aligned}
\tag{A}
$$

are rewritten in such a way that $x_1, x_2, x_3, \ldots$ are positive and that the values $y_1/x_1, y_2/x_2, \ldots$ of β, given by these equations on the assumption that $\epsilon_1, \epsilon_2, \ldots$ are zero, form a decreasing sequence. The procedure in question consists of choosing the rth equation of condition, such that we have

$$
\begin{aligned}
x_1 + x_2 \cdots + x_{r-1} &< x_r + x_{r+1} \cdots + x_n, \\
x_1 + x_2 \cdots + x_r &> x_{r+1} + x_{r+2} \cdots + x_n,
\end{aligned}
\tag{B}
$$

and to set

$$
\beta = y_r/x_r.
$$

This value of β *minimizes* the sum of all the absolute deviations of the other values. For, if $\epsilon_1, \epsilon_2, \ldots$ denote these deviations, then $\epsilon_1, \epsilon_2 \ldots \epsilon_{r-1}$ will be positive and $\epsilon_{r+1}, \epsilon_{r+2} \ldots \epsilon_n$ will be negative. If we increase the above value of β by the infinitesimal quantity $\delta\beta$, the sum of the positive deviations $\epsilon_1, \epsilon_2, \ldots$ will decrease by

$$\delta\beta(x_1 + x_2 \cdots + x_{r-1}),$$

but the sum of the absolute values of the negative deviations will increase by

$$\delta\beta(x_{r+1} + x_{r+2} \cdots + x_n),$$

and the deviation ϵ_r will become $\delta\beta.x_r$. The sum of the absolute values of all the deviations will therefore be increased by

$$\delta\beta(x_r + x_{r+1} \cdots + x_n - x_1 - x_2 \cdots x_{r-1}).$$

By the conditions determining the choice of the rth equation, this quantity is positive. Likewise, we see that if y_r/x_r is decreased by $\delta\beta$, then the sum of the absolute deviations will be increased by the positive quantity

$$\delta\beta(x_1 + x_2 \cdots + x_r - x_{r+1} - x_{r+2} \cdots - x_n).$$

Thus for the two cases of an increase or a decrease of the value y_r/x_r of β, the sum of the absolute deviations is increased. This argument seems to give a big advantage to the above value of β, a value which reduces to the median for the problem of finding a mean for an odd number of observations. But only the calculus of probabilities enables us to assess this advantage and I will therefore apply it to this difficult question.

The only information we will be using is (a) that the equation of condition

$$0 = y_r + \beta x_r - \epsilon_r$$

gives, upon disregarding the errors, a value of β smaller than do the $r - 1$ preceding equations and larger than do the $n - r$ subsequent equations, and (b) that we have

$$x_1 + x_2 \cdots + x_{r-1} < x_r + x_{r+1} \cdots + x_n,$$
$$x_1 + x_2 \cdots + x_r > x_{r+1} + x_{r+2} \cdots + x_n.$$

We have

$$\beta = \frac{y_1}{x_1} - \frac{\epsilon_1}{x_1} = \frac{y_r}{x_r} - \frac{\epsilon_r}{x_r}$$

which gives

$$\frac{\epsilon_1}{x_1} = \frac{y_1}{x_1} - \frac{y_r}{x_r} + \frac{\epsilon_r}{x_r}.$$

Thus, since y_1/x_1 exceeds y_r/x_r, ϵ_1/x_1 exceeds ϵ_r/x_r. The same holds for $\epsilon_2/x_2, \epsilon_3/x_3 \ldots$ up to ϵ_{r-1}/x_{r-1}. In the same way, it will be seen that $\epsilon_{r+1}/x_{r+1}, \epsilon_{r+2}/x_{r+2} \ldots \epsilon_n/x_n$ are less than ϵ_r/x_r. Thus the only conditions to which we will subject the errors and the equations of condition are the following.

$$\left\{ \begin{array}{c} s > r \\ \dfrac{\epsilon_s}{x_s} < \dfrac{\epsilon_r}{x_r} \end{array} \right\}, \quad \left\{ \begin{array}{c} s < r \\ \dfrac{\epsilon_s}{x_s} > \dfrac{\epsilon_r}{x_r} \end{array} \right\} \qquad (c)$$

$$x_1 + x_2 \cdots + x_{r-1} < x_r + x_{r+1} \cdots + x_n,$$
$$x_1 + x_2 \cdots + x_r > x_{r+1} + x_{r+2} \cdots + x_n.$$

We will determine the probability distribution of the error ϵ_r purely from these conditions. Moreover, we will pay no attention to the order of occurrence of the first $r - 1$ equations of condition, or the last $n - r$, or to the values of the quantities $y_1, y_2, \ldots, y_n$.

As above, let us represent by $\phi(\epsilon)$ the probability law of the error ϵ of the observations. Also to express that this probability is the same for positive and negative errors, let us assume that $\phi(\epsilon)$ is a function of ϵ^2.

Now, if we assume ϵ_r to be positive, the probability that ϵ_1 exceeds $x_1.\epsilon_r/x_r$ is

$$\frac{1}{2} - \frac{\frac{1}{2} \int d\epsilon.\phi(\epsilon)}{k},$$

where the integral $\int d\epsilon.\phi(\epsilon)$ is taken from $\epsilon = 0$ to $\epsilon = x_1.\epsilon_r/x_r$, and k is, as above, this integral taken from 0 to ∞. The probability that the quantities $\epsilon_1/x_1, \epsilon_2/x_2 \ldots \epsilon_{r-1}/x_{r-1}$ are all larger than ϵ_r/x_r is therefore proportional to the product of the $r - 1$ factors

$$1 - \frac{\int d\epsilon.\phi(\epsilon)}{k}, \quad 1 - \frac{\int d\epsilon.\phi(\epsilon)}{k}, \ldots,$$

the integral in the first factor being taken from $\epsilon = 0$ to $\epsilon = x_1.\epsilon_r/x_r$, the integral in the second being taken from $\epsilon = 0$ to $\epsilon = x_2.\epsilon_r/x_r$, and so on.

Similarly, since all the quantities $\epsilon_{r+1}/x_{r+1}, \epsilon_{r+2}/x_{r+2} \ldots \epsilon_n/x_n$ are assumed to be smaller than ϵ_r/x_r, we see by the same reasoning that probability of this assumption is proportional to the product of the $n - r$ factors

$$1 + \frac{\int d\epsilon.\phi(\epsilon)}{k}, \quad 1 + \frac{\int d\epsilon.\phi(\epsilon)}{k}, \ldots.$$

The integral in the first factor is taken from $\epsilon = 0$ to $\epsilon = x_{r+1}.\epsilon_r/x_r$, that in the second factor is taken from $\epsilon = 0$ to $\epsilon = x_{r+2}.\epsilon_r/x_r$, and so on. The probability of the error ϵ_r is $\phi(\epsilon_r)$. Thus the probability that the error of the rth observation is ϵ_r and that the value of β given by the rth equation is smaller than the values given by the preceding equations and exceeds the values of the subsequent equations, this probability, I say, is proportional to the product of the $n - 1$ above factors and $\phi(\epsilon_r)$.

ϵ being assumed to be very small, we have to order ϵ^2

$$\int d\epsilon.\phi(\epsilon) = \epsilon.\phi(0) + \frac{1}{2} \epsilon^2.\phi'(0),$$

where $\phi'(0)$ is the value of $d\phi(\epsilon)/d\epsilon$ for $\epsilon = 0$. In the present situation we have $\phi'(0) = 0$, since $\phi(\epsilon)$ is a function of ϵ^2, so that

$$\int d\epsilon.\phi(\epsilon) = \epsilon\phi(0).$$

The above factors therefore become, on setting $\epsilon_r/x_r = \zeta$,

$$1 - x_1\zeta.\frac{\phi(0)}{k} \, , \, 1 - x_2\zeta.\frac{\phi(0)}{k} \, , \ldots \, 1 - x_{r-1}\zeta.\frac{\phi(0)}{k} \, ,$$

$$1 + x_{r+1}\zeta.\frac{\phi(0)}{k} \, , \ldots 1 + x_n\zeta.\frac{\phi(0)}{k}.$$

If we denote by $\phi''(0)$ the value of $d^2\phi(\epsilon)/d\epsilon^2$ when $\epsilon = 0$, $\phi(\epsilon_r)$ becomes

$$\phi(0) + \frac{1}{2}x_r^2.\zeta^2.\phi''(0).$$

The sum of the natural logarithms of all these factors, on dividing the factor $\phi(\epsilon_r)$ by $\phi(0)$, is to order ζ^2

$$- \quad \zeta.\frac{\phi(0)}{k}.(x_1 + x_2 \cdots + x_{r-1} - x_{r+1} - x_{r+2} - \cdots - x_n)$$

$$- \quad \frac{\zeta^2}{2}.\left(\frac{\phi(0)}{k}\right)^2.(x_1^2 + x_2^2 \cdots + x_r^2 + x_{r+1}^2 \cdots + x_n^2)$$

$$+ \quad \frac{1}{2}.x_r^2.\zeta^2\left[\frac{\phi''(0)}{k} + \left(\frac{\phi(0)}{k}\right)^2\right].$$

The probability of ζ is therefore proportional to the base c of natural logarithms, raised to a power given by the above expression. We must note that in view of the conditions imposed on the choice of the rth equation, the quantity

$$x_1 + x_2 \cdots + x_{r-1} - x_{r+1} - x_{r+2} - \cdots - x_n$$

is, disregarding the sign, a quantity less than x_r. Thus, assuming ζ to be of order $1/\sqrt{n}$, the number of observations being assumed very large, the term depending on the first power of ζ in the above function is of order $1/\sqrt{n}$. We can therefore neglect it as well as the last term of this function. If therefore we let $S.x_i^2$ denote the entire sum

$$x_1^2 + x_2^2 \cdots + x_n^2,$$

the probability of ζ is proportional to

$$c^{-\frac{\zeta^2}{2}.\left(\frac{\phi(0)}{k}\right)^2.S.x_i^2}, \qquad \text{(C)}$$

where ζ or ϵ_r/x_r is the error of the value y_r/x_r given for β by the rth equation. The value given by the most advantageous method is, as in the previous section,

$$\beta = \frac{S.x_i y_i}{S.x_i^2}.$$

The probability of an error ζ in this result is proportional to

$$c^{-\dfrac{k}{2k''}\zeta^2.S.x_i^2}, \tag{D}$$

where k'' is always the integral $\int \epsilon^2 d\epsilon.\phi(\epsilon)$, taken from $\epsilon - 0$ to $\epsilon = \infty$. The result of the method we have just examined, and which we will call the method of *situation*, will be preferable to that of the most advantageous method if the coefficient of $-\zeta^2$, pertaining to it, exceeds the coefficient pertaining to the most advantageous method; for then the law of the probability of errors will be more rapidly decreasing. Thus the method of situation must be preferred if

$$\left(\dfrac{\phi(0)}{k}\right)^2 > \dfrac{k}{k''}. \tag{E}$$

In the opposite case, the most advantageous method is preferable. If, for example, we have

$$\phi(x) = e^{-hx^2},$$

then k becomes $\sqrt{\pi}/2\sqrt{h}$ and k'' becomes $\sqrt{\pi}/4h\sqrt{h}$, which gives $k/k'' = 2h$. The term $(\phi(0)/k)^2$ becomes $4h/\pi$, so that we have $2h > 4h/\pi$. The most advantageous method must therefore be preferred.

Combining the results of these two methods, we can obtain a result for which the law of the probability of errors is more rapidly decreasing. Let us always call ζ the error of the result of the method of situation, and denote by ζ' the error of the result of the most advantageous method. The first of these results is, as we have seen, y_r/x_r, and the second is $S.x_iy_i/S.x_i^2$. If we denote $S.x_i\epsilon_i$ by ℓ, then $\ell/S.x_i^2$ is the error of this last result. Thus we have $\ell = \zeta'.S.x_i^2$. The joint probability of ℓ and ζ is by Section 21 of the second book of the *Analytical Theory* proportional to

$$\int d\omega.c^{-\ell\omega\sqrt{-1}}\,\phi(x_r\zeta).c^{x_r\zeta.\omega\sqrt{-1}} \begin{cases} \int d\epsilon.\phi(\epsilon).c^{x_1\epsilon\omega\sqrt{-1}} \\ \times \int d\epsilon.\phi(\epsilon).c^{x_2\epsilon\omega\sqrt{-1}} \\ \times \text{ etc.,} \end{cases} \tag{F}$$

where the integral w.r.t. ω runs from $\omega = -\pi$ to $\omega = \pi$. The integral w.r.t. ϵ in the factor $\int d\epsilon.\phi(\epsilon).c^{x_1\epsilon\omega\sqrt{-1}}$ must be taken, by the above, from $\epsilon = x_1\zeta$ to $\epsilon = \infty$. Expanding this factor in powers of ω gives

$$\int d\epsilon.\phi(\epsilon) + x_1\omega\sqrt{-1}\int \epsilon d\epsilon.\phi(\epsilon) - x_i^2.\dfrac{\omega^2}{2}\int \epsilon^2 d\epsilon.\phi(\epsilon)\ldots.$$

Taking the integral between the above limits, we have, to order ζ^2

$$\int d\epsilon.\phi(\epsilon) = k - x_1\zeta.\phi(0).$$

Similarly, neglecting terms of order $\zeta^2\omega, \zeta^3\omega^2, \ldots$, we have

$$x_1\omega\sqrt{-1}. \int \epsilon d\epsilon.\phi(\epsilon) = k'.x_1\omega.\sqrt{-1},$$

$$-\frac{x_1^2}{2}.\omega^2. \int \epsilon^2 d\epsilon.\phi(\epsilon) = -\frac{k''}{2}.x_1^2\omega^2,$$

where k' is the integral $\int \epsilon d\epsilon.\phi(\epsilon)$ taken from $\epsilon = 0$ to $\epsilon = \infty$. Neglecting ω^3, the factor in question is therefore, according to the analysis of the cited section of the second book,

$$k - x_1\zeta.\phi(0) + k'.x_1\omega\sqrt{-1} - \frac{k''}{2}.x_1^2\omega^2.$$

Its natural logarithm is

$$-x_1\zeta.\frac{\phi(0)}{k} + \frac{k'}{k}.x_1\omega\sqrt{-1} - \frac{k''}{2k}.x_1^2\omega^2$$
$$-\frac{x_1^2}{2}.\left(\zeta\frac{\phi(0)}{k} - \frac{k'}{k}.\omega\sqrt{-1}\right)^2 + \log.k.$$

Successively changing x_1 to $x_2, x_3, \ldots x_{r-1}$, we will have the logarithms of the subsequent factors, up to the factor pertaining to x_{r-1}.

For the factor $\int d\epsilon.\phi(\epsilon).c^{x_{r+1}\epsilon\omega\sqrt{-1}}$ the integral must be taken from $\epsilon = -\infty$ to $\epsilon = x_{r+1}.\zeta$. Since $\int \epsilon d\epsilon.\phi(\epsilon)$ becomes $-k'$, the logarithm of this factor is

$$x_{r+1}.\zeta.\frac{\phi(0)}{k} - \frac{k'}{k}.x_{r+1}.\omega\sqrt{-1} - \frac{k''}{2k}.x_{r+1}^2\omega^2$$
$$-\frac{x_{r+1}^2}{2}.\left(\zeta\frac{\phi(0)}{k} - \frac{k'}{k}\omega\sqrt{-1}\right)^2 + \log.k.$$

We will have the logarithms of the subsequent factors on changing x_{r+1} successively to $x_{r+2}, x_{r+3} \ldots x_n$. The factor $\phi(x_r\zeta).c^{x_r\zeta\omega\sqrt{-1}}$ is equal to

$$\left[\phi(0) + \frac{x_r^2.\zeta^2}{2}\phi''(0)\right].c^{x_r\zeta\omega\sqrt{-1}},$$

and its logarithm is

$$\frac{x_r^2}{2}.\zeta^2.\frac{\phi''(0)}{\phi(0)} + x_r\zeta\omega\sqrt{-1} + \log.\phi(0).$$

Now if we collect all the logarithms, if we then consider the conditions (c) which the rth equation satisfies, and finally if we return from logarithms to natural numbers, we find, upon neglecting what it is permissible to neglect, that the joint probability of ℓ and ζ is proportional to

$$\int d\omega.c^{\displaystyle -\ell\omega.\sqrt{-1} - \left[\left(\zeta\frac{\phi(0)}{k} - \frac{k'}{k}\omega.\sqrt{-1}\right)^2 + \frac{k''}{k}.\omega^2\right].\frac{S.x_i^2}{2}}.$$

Upon setting therefore

$$F = \left(\frac{k''}{k} - \frac{k'^2}{k^2}\right) \cdot \frac{S.x_i^2}{2},$$

the joint probability of ζ and ζ' will be proportional to

$$c^{-\frac{\zeta^2}{2} \cdot \left(\frac{\phi(0)}{k}\right)^2 \cdot S.x_i^2 - \frac{\left(\zeta' - \zeta \cdot \frac{k'}{k} \cdot \frac{\phi(0)}{k}\right)^2 \cdot (S.x_i^2)^2}{4F}}$$

$$\times \int dw.c^{-F.\left\{w + \frac{\left(\zeta' - \zeta \cdot \frac{k'}{k} \cdot \frac{\phi(0)}{k}\right)\sqrt{-1}.S.x_i^2}{2F}\right\}^2}.$$

By the analysis of Section 21 of the second book, the integral w.r.t. w may be taken from $w = -\infty$ to $w = +\infty$. Then the above probability becomes proportional to

$$c^{-\frac{\zeta^2}{2}.S.x_i^2 \cdot \left(\frac{\phi(0)}{k}\right)^2 - \frac{\left(\zeta' - \zeta \cdot \frac{k'}{k} - \frac{\phi(0)}{k}\right)^2}{2\left(\frac{k''}{k} - \frac{k'^2}{k^2}\right)}.S.x_i^2},$$

an expression that may also be put in the form

$$c^{-\frac{k}{2k''} \cdot \zeta'^2 .S.x_i^2 - \frac{k''}{k} \cdot \frac{\left(\zeta \cdot \frac{\phi(0)}{k} - \zeta' \frac{k'}{k''}\right)^2}{2\left(\frac{k''}{k} - \frac{k'^2}{k^2}\right)}.S.x_i^2}.$$

If we call e the excess of the value of β given by the most advantageous method, over that given by the method of situation, then $\zeta = \zeta' - e$. Let us suppose

$$\zeta' = u + \frac{e.\frac{\phi(0)}{k} \cdot \left(\frac{\phi(0)}{k} - \frac{k'}{k''}\right)}{\frac{k}{k''} - \frac{k'^2}{k''^2} + \left(\frac{\phi(0)}{k} - \frac{k'}{k''}\right)^2}.$$

The probability of u will be proportional to

$$c^{-\frac{u^2}{2}.S.x_i^2 \left\{\frac{k}{k''} + \frac{\frac{k''}{k}\left(\frac{\phi(0)}{k} - \frac{k'}{k''}\right)^2}{\frac{k''}{k} - \frac{k'^2}{k^2}}\right\}}.$$

The result of the most advantageous method must therefore be decreased by the amount

$$\frac{e.\frac{\phi(0)}{k} \cdot \left(\frac{\phi(0)}{k} - \frac{k'}{k''}\right)}{\frac{k}{k''} - \frac{k'^2}{k''^2} + \left(\frac{\psi(0)}{k} - \frac{k'}{k''}\right)^2}.$$

and the probability of an error u in this result so corrected will be proportional to the preceding exponential. The weight of the new result will be increased if

$$\frac{\phi(0)}{k} - \frac{k'}{k''} \neq 0.$$

There is therefore an advantage in correcting the result of the most advantageous method in this way. Our ignorance of the law of probability of the errors of observation render this correction impractical. But it is remarkable that in the case where this probability is proportional to c^{-hx^2}, i.e., where we have $\phi(x) = c^{-hx^2}$, the quantity $\frac{\phi(0)}{k} - \frac{k'}{k''}$ is zero. Then the result of the most advantageous method receives no correction from the result of the method of situation, and the law of probability of errors remains unchanged.

The Logistic Growth Curve

Comments on Verhulst (1845)

1 Introduction

Interest in a better understanding of the growth of populations was undoubtedly raised by Malthus (1798) in his famous *Essay on the Principle of Population*. Using hardly any mathematics, Malthus claimed that "population, when unchecked, increases in a geometrical ratio and subsistence in an arithmetical ratio." It was in this spirit that the Belgian mathematician Pierre-François Verhulst (1804–1849), with some encouragement from his compatriot Adolphe Quetelet, studied a model that included a term slowing population growth.

Specifically, Verhulst supposes that the initial unchecked growth of the population is described by

$$\frac{M\,dN}{N\,dt} = l \, , \tag{a}$$

where N is the population size at time t, l is an unknown constant, and $M = \log_{10} e$. (Verhulst needs M since he takes his logarithms to base 10.) But when N reaches size N_0, he replaces the exponential growth corresponding to (a) by

$$\frac{M\,dN}{N\,dt} = m - nN \, ,$$

where m and n are unknown constants and time is now measured from the moment when $N = N_0$. This leads to Verhulst's eq. (4), which gives N in inverse form. Inverting (4), we may write

$$N = \frac{\alpha e^{\alpha t}}{\beta e^{\alpha t} + \alpha/N_0 - \beta} \, , \tag{b}$$

where $\alpha = m/M$ and $\beta = n/M$. Essentially, Verhulst here names the right side of (b) the *logistic curve*. From (b) we see that N increases from N_0 at $t = 0$ to $\alpha/\beta = m/n$ at $t = \infty$, corresponding in current terminology to a logistic cdf truncated on the left, as in the figure.

Verhulst goes on to obtain properties of the logistic curve and in eq. (B) of Section 6 even considers an asymmetric generalization. Note that $\frac{1}{\mu-1\sqrt{\mu}} = \mu^{-1/(\mu-1)}$.

We truncate the paper (on the right) at this point. Verhulst proceeds to some crude parameter estimation and applies his results to a detailed examination of the populations of Belgium and France. If conditions remain stable, he predicts respective maxima of 6.6 and 40 million. The population of Belgium was 4.8 million in 1845 and that of France 34.4 million in 1842. Of course, conditions did change, in the first instance because of the industrial revolution. Recent figures are: Belgium 10.2 and France 59.0 million.

2   U.S. Population and References

Surprisingly, Verhulst uses the distant U.S. decennial censuses for 1790 to 1840 to illustrate exponential increase when there are no checks on growth. Interpolating just linearly, he constructs Table (A) giving population figures at 5-year intervals and showing that the population doubled about every 25 years. The results are striking, but his remark following the table can only be described as quaint.

From subsequent census figures, as, e.g., in *The World Almanac and Book of Facts 2000*, we easily find that this rate of doubling continues until 1860. After this date there is a clear but not altogether regular decline in the ratio r, which reaches 1.258 for 1990. The population has taken more than 60 years to double from its 123.2 million in 1930, being 240.8 million in 1990.

Many more detailed studies of the growth of the U.S. population have, of course, been made. Pearl and Reed (1920) rediscovered the logistic distribution in this context and wrote extensively on it. Initially unaware of Verhulst's work, they remained reluctant to give credit to Verhulst. In contrast, Yule (1925) does so and appears to have introduced the term "logistic curve" into English.

A detailed multiauthored treatment of the logistic distribution is given in Balakrishnan (1991). We must also point out that the logistic distribution was in fact (but not in name) already introduced in Verhulst (1838). An even earlier occurrence of the logistic, as a limiting distribution (Poisson, 1824), is noted by Hald (1998). For a biographical account of Verhulst see Miner (1933).

3   Specific Notes

For greater ease of reading Verhulst's symbols p, π, b are replaced in the translation by N, N_0', and N_0, respectively.

The figure, absent from our copy of Verhulst's paper, was reconstructed from the text, and drawn by Norma Elwick.

References

Balakrishnan, N. (ed.) (1991). *Handbook of the Logistic Distribution*. Marcel Dekker, New York.

Hald, A. (1998). *A History of Mathematical Statistics from 1750 to 1930*. Wiley, New York.

Malthus, T.R. (1798). An essay on the principle of population as it affects the future improvement of society, with remarks on the speculations of Mr. Godwin, M. Condorcet, and other writers. J. Johnson, London.

Miner, J.R. (1933). Pierre-Francois Verhulst, the discoverer of the logistic curve. *Human Biol.*, **5**, 673–689.

Pearl, R. and Reed, L.J. (1920). On the rate of growth of the population of the United States since 1790 and its mathematical representation. *Proc. Nat. Acad. Sci.*, **6**, 275–288.

Poisson, S.D. (1829). Suite du Mémoire sur la probabilité du résultat moyen des observations, inséré dans la Connaissance des Tems de l'année 1827. *Connaiss. Tems* pour 1832, pp. 3–22.

Verhulst, P.F. (1838). Notice sur la loi que la population suit dans son accroissement. *Corresp. Math. Phys. Publ.* par A. Quetelet (Brussels), **10**, 113–121.

Verhulst, P.F. (1845). Recherches mathématiques sur la loi d'acroissement de la population. Nouveaux mémoires de l'académie royale des sciences et belles-lettres de Bruxelles, **18**, 1–38.

Yule, G.U. (1925). The growth of population and the factors which control it (with discussion). *J. Roy. Statist. Soc.*, **88**, 1–62.

Mathematical Investigations on the Law of Population Growth

P.-F. Verhulst

General Theory

1

Of all the problems that political economy offers for the consideration of philosophers, one of the most interesting is doubtless the knowledge of the law regulating the progression of the population. To resolve this exactly it would be necessary to assess the influence of the numerous causes that impede or favor the multiplication of the human species. And since several of these causes are variable by nature and by their mode of action, the problem considered in all its generality is clearly insoluble.

However, we must observe that to the extent that civilization perfects itself, the influence of purely perturbatory causes declines more and more, to let constant causes dominate. Thus it becomes permissible at a certain period to ignore the former, but to regard the data in question as subject to slight variations.

Consequently, in order to quantify the population principle we will begin by ignoring accidental causes, but are nevertheless far from denying their importance in the current state of society. If statistical data allowed the same precision as data in the experimental sciences, such as physics and chemistry, we could assess the influence of the neglected causes by comparing the results of calculation with the actual observations. But unfortunately statistics is still too new a science for us to have complete confidence in the figures that it furnishes.

2

Among the causes exerting a constant action on population growth, we will set down the fertility of the human race, the health conditions of the country, the customs of the nation under consideration, and its civil and religious laws. The variable causes that we cannot regard as accidental are generally summed up in the increasing difficulty experienced by the population in procuring its means of subsistence when it has become so numerous that all the good land is occupied.

If one does not take into account the difficulty just mentioned, it must be granted that, by virtue of the constant causes, the population must increase in geometric progression. Actually, if 1000 souls have become 2000 at the end of 25 years, for example, there is no reason why these 2000 should not become 4000 at the end of the following 25 years.

The United States provide us with an example of such a rapid rate of population growth. According to the U.S. official censuses, the counts were

In 1790.	. . . .	3,929,827	souls,
1800.	. . . .	5,305,925	
1810.	. . . .	7,239,814	
1820.	. . . .	9,638,151	
1830.	. . . .	12,866,020	
1840.	. . . .	17,062,566	

If we take for the population of 1795 the figure 4,617,876, the average of the figures for 1790 and 1800, and if the same is done for the years 1805, 1815, 1825, and 1835, we can evaluate approximately the progress of the population in 5-year intervals. It is in this manner that we have formed the following table, in which we have rounded the figures and denoted by r the ratio of each population to that 25 years earlier.

Years	Population	Values of r	
1790.	3,930,000		
1795.	4,618,000		
1800.	5,306,000		
1805.	6,273,000		
1810.	7,240,000		(A)
1815.	8,439,000	2.147	
1820.	9,638,000	2.087	
1825.	11,252,000	2.120	
1830.	12,866,000	2.052	
1835.	14,964,000	2.076	
1840.	17,063,000	2.021	

We have not taken immigration into account because we regard this as largely compensated by the obstacles that slavery puts in the way of the multiplication of blacks, in the States of the South.

3

Let N denote the population size, t the time, and k and l arbitrary constants. If the population increases in geometric progression while the time increases in arithmetic progression, we have the following relation between these two quantities:

$$N = k10^{lt}.$$

Let N' be the population size at time t'. It follows that

$$N = N'10^{l(t-t')}$$

and if we denote by N'_0 the population size when we begin measuring time, the preceding equation becomes

$$N = N'_0 10^{lt}. \tag{1}$$

Under the geometric series hypothesis, the population curve is therefore an *exponential* [une *logarithmique*], in which the ordinates and the abscissae represent, respectively, the population sizes and the times elapsed, the ordinate being N_0 at the origin.

Twice differentiating equation (1) and denoting by M the factor by which natural logarithms must be multiplied to convert them to common logarithms, we have

$$\frac{d^2 N}{dt^2} = \frac{N'_0 l^2 10^{lt}}{M^2} = \frac{l^2 N}{M^2},$$

which shows that the curve always turns its convexity towards [i.e., is always convex as seen from] the horizontal axis.

The Malthusian period of 25 years assumes that N becomes $2N$ as t becomes $t + 25$, the year being taken as unit of time. We therefore have the equations

$$\begin{aligned} 2N &= N'_0 10^{lt+25l}, \\ 2N &= 2N'_0 10^{lt}; \end{aligned}$$

from which $2 = 10^{25l}$ and

$$l = \frac{1}{25} \log 2 = 0.012041200.$$

We will no longer insist on the geometric progression hypothesis, seeing that it holds only in quite exceptional circumstances, for example, when a territory, fertile and virtually unlimited in extent, is inhabited by a people with a very advanced civilization, such as the first settlers of the United States.

4

Differentiation of equation (1) gives

$$\frac{M\,dN}{p\,dt} = l.$$

Since this quantity is constant, it may be taken as a measure of the energy with which the population tends to develop when it is not held back by the fear of lack of provisions. We have also, with increasing accuracy the smaller ΔN and Δt,

$$M\Delta N = lN\Delta t,$$

and if we take Δt as a one-year interval

$$\frac{\Delta N}{N} = \frac{l}{M}.$$

In the case of the geometric progression, this means that the annual excess of births over deaths, divided by the corresponding population size, is a constant ratio. In terms of purely numerical results, it does not matter whether the number of births increases or decreases, provided that the number of deaths increases or decreases by the same amount. If we notice therefore that the coefficient l/M is declining, we are in the same way able to attribute the cause to a decline in the fertility of the population or to an increase in its mortality, i.e., to preventive obstacles or to destructive obstacles.

It is an observed fact that in all of Europe the ratio of the annual excess of births over deaths to the corresponding population size, and consequently the coefficient l/M, is constantly declining. But the annual growth, whose absolute value continually increases when there is a geometric progression, appears to follow a progression that is arithmetic at the very outside. This remark confirms the famous aphorism of Malthus, that *the population tends to grow in geometric progression whereas the production of subsistence follows an arithmetic progression at the very best*, since the population is obliged to adjust itself to its means of subsistence.

We can make an infinity of hypotheses about the law of decline of the coefficient l/M. The simplest consists in regarding this decline as proportional to the growth of the population, from the moment when the difficulty of finding good land has made itself felt. The population at that special moment, from which we now measure time, will be called a *normal population* and designated N_0. Then, with n denoting an arbitrary coefficient, we replace the differential equation $\frac{M\,dN}{N\,dt} = l$, connected with the geometric progression, by

$$\frac{M\,dN}{N\,dt} = l - n(N - N_0). \tag{2}$$

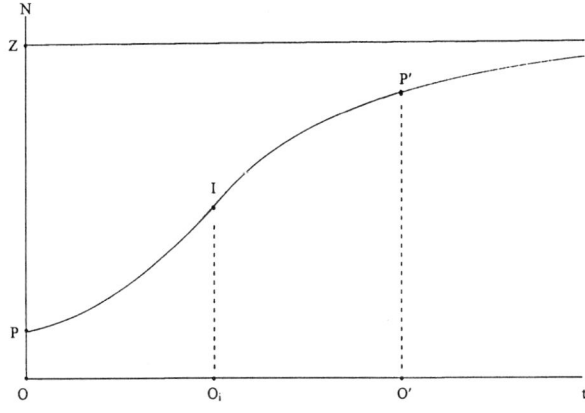

FIGURE 1.

Putting $m = l + nN_0$ for brevity, we have

$$\frac{M\,dN}{N\,dt} = m - nN$$

and

$$dt = \frac{M\,dN}{mN - nN^2}. \tag{3}$$

Since $t = 0$ corresponds to $N = N_0$, integration gives

$$t = \frac{1}{m} \log\left[\frac{N(m - nN_0)}{N_0(m - nN)}\right]. \tag{4}$$

We will give the name *logistic* [*logistique*] to the curve (see the figure) determined by the preceding equation. The curve is seen to have an asymptote parallel to the horizontal axis, at a distance m/n from the origin, for $N = m/n$ corresponds to $t = \infty$. This value of N is equal to the ordinate OZ, which represents the extreme limit of the population.

Differentiation of equation (2) gives

$$M^2 \frac{d^2N}{dt^2} = (m - 2nN)(mN - nN^2),$$

which shows that the curve has a point of inflection I at $N = \frac{1}{2} \cdot \frac{m}{n}$. It is there that the curve changes from convex to concave towards the horizontal axis.

5

Equation (4) takes a simpler form if the origin is moved to the point O_i, the foot of the ordinate N_i at the point of inflection. This may be done by changing t to $t + t_i$, where t_i denotes the abscissa OO_i. If we note that

$$t_i = \frac{1}{m} \log \left[\frac{N_i(m - nN_0)}{N_0(m - nN_i)} \right]$$

and that $N_i = \frac{m}{2n}$, then

$$t = \frac{1}{m} \log \left\{ \frac{N}{\frac{m}{n} - N} \right\}. \tag{5}$$

This equation is such that if N is changed to $\frac{m}{n} - N$, t simply changes sign. Hence results the property that *the sum of the ordinates situated at the same distance from the point of inflection is constant and equal to the limit ordinate $\frac{m}{n}$.*

Let us take $O_iO' = OO_i = \frac{1}{m} \log \left\{ \frac{N_0}{\frac{m}{n} - N_0} \right\}$. In view of the above property we have

$$O'P' = \frac{m}{n} - N_0,$$

i.e., *when the population has reached an age corresponding to the point O', it cannot grow by more than the size of the normal population.*

Consideration of the three special points O, O_i, and O', leads us to divide the infinite extent of time into four *periods*, to each of which it would be easy to attach a distinctive character borrowed either from agriculture or political economy. In the first period, the population curve is an exponential which changes to a logistic at the point P; during this period only good land is cultivated. In the third period, which is of the same duration as the second, only poor land is brought under cultivation. Finally, in the fourth period, the population increases only on account of improvements in already cultivated soil or the importation of foreign trade. Having said this, we believe it superfluous to observe that these connections between the progress of the population and the expansion of agriculture are not capable of great precision.

6

Since at the point of inflection we have $d^2N = 0$, the differences ΔN are essentially equal in the neighborhood of this point; i.e., the ordinates increase in arithmetic progression. The ordinate IO_i, being half the ordinate at $t = \infty$, it follows that *the population tends eventually to become double its size at the time when the annual increase is constant.*

On assuming

$$\frac{M\,dN}{dt} = mN - nN^\mu \tag{B}$$

we would find that the ordinate at the point of inflection is the $\frac{1}{\mu - 1\sqrt{\mu}}$th part of the ordinate at $t = \infty$. In general, *the more rapidly increasing* *the function representing the constraint experienced by the population, the larger the ordinate at the point of inflection relative to the ordinate at $t = \infty$.* To show this, we take

$$\frac{M\,dN}{N\,dt} = m - f(N)$$

for the differential equation of the population curve. The ordinate at ∞, N_∞, follows from the equation

$$m - f(N_\infty) = 0,$$

whereas the ordinate N_i at the point of inflection is given by the equation

$$f(N_\infty) = f(N_i) + N_i f'(N_i),$$

f' denoting the derivative of the function f. From this we have

$$\frac{f(N_\infty)}{f(N_i)} = 1 + N_i \frac{f'(N_i)}{f(N_i)}.$$

Now, the more rapidly increasing the function $f(N)$, the larger is $f'(N_i)$ relative to $f(N_i)$, and the larger is $f'(N_i)/f(N_i)$ relative to N_i; and because of the last equation, the larger is $f(N_\infty)$ relative to $f(N_i)$ or N_i. Hence N_∞, being the function inverse to $f(N_\infty)$, is the smaller relative to N_i.

Goodness-of-Fit Statistics

Comments on Abbe (1863)

1   Introduction

Kendall (1971) gives an excellent concise account of Abbe's 1863 paper, especially of the ingenious mathematical techniques used. Naturally he can not, in limited space, capture the full flavor of the paper.

In an introduction, not translated here, Abbe considers various questions of goodness of fit. He expresses the need to examine the residuals resulting from the method of least squares, as developed by Gauss. Are these residuals random and in accord with the degree of variability to be expected from the observational process in question? If the residuals have too high a mean square or if their successive differences show a noticeable trend, then we must conclude that either the assumed functional relationship is incorrect or that there are systematic errors; and this with increasing confidence as the number of observations becomes larger.

Although these issues are often settled by inspection, Abbe continues, it seems desirable to have more refined methods. Assuming, as Gauss did in his "first proof" of the method of least squares (Gauss, 1809), that the residuals X_j $(j = 1, \ldots, n)$ are independently drawn from the normal distribution (1), Abbe is motivated to find the exact distribution of $\Delta = \sum_1^n X_j^2$ and of $\sum_1^{n-1}(X_j - X_{j+1})^2 = \Theta^*$ (say). In the case of Δ he claims no priority, but gives no names. One who anticipated him was Bienaymé (1852) who, however, came upon the distribution incidentally, in the course of evaluating a related integral (Lancaster, 1966; David, 1998).

To simplify the second derivation, Abbe adds the circularizing term $(X_n - X_1)^2$, giving his Θ. Helmert, who was unaware of Abbe's work earlier, considers Θ^* primarily asymptotically in Helmert (1907). Even von Neumann et al. (1941) can handle the distribution of Θ^* only by a moment approximation (except for $n = 3$).

Finally, Abbe obtains the distribution of $\mu = \Theta/\Delta$, essentially the first circular serial correlation coefficient for a known mean. Helmert (1907) calls this Abbe's Criterion and proposes a decircularized version. The distribution of the first circular serial correlation coefficient for an unknown mean was not obtained until Anderson (1942) who was unaware of Abbe's work. See also Hald (1998, p. 637).

2 Mathematical Techniques

Abbe marshals an impressive array of mathematical techniques and facts. His most important tool is clearly contour integration introduced by Cauchy, mainly in the 1820s. "Discontinuity factors," so named by Dirichlet, also originated in this period (Hald, 1998). An appropriate discontinuity factor was often used in the nineteenth century, as here by Abbe, to transform an inconvenient region of integration into a more convenient one. He also makes considerable use of the theory of determinants, including circulants, in connection with the diagonalization of the quadratic form Θ.

The "well-known formula" referred to towards the end of Section 3 is seen from the footnote reference to be

$$\frac{\sin nx}{\sin 2x} = 2^{\frac{n-2}{2}} \left(\cos 2x - \cos \frac{2\pi}{n} \right) \left(\cos 2x - \cos \frac{4\pi}{n} \right) \cdots$$
$$\cdots \left(\cos 2x - \cos \frac{(n-2)\pi}{n} \right) \quad (n \text{ even})$$

$$\frac{\sin nx}{\sin 2x} = 2^{\frac{n-1}{2}} \left(\cos 2x - \cos \frac{2\pi}{n} \right) \left(\cos 2x - \cos \frac{4\pi}{n} \right) \cdots$$
$$\cdots \left(\cos 2x - \cos \frac{(n-1)\pi}{n} \right) \quad (n \text{ odd}).$$

3 Conclusion

It is only in the very last sentence of Abbe's treatise that the 23-year-old author shares a typical youthful failing of the "space does not permit ..." variety. As far as is known, he did not return to the subject of this paper, as Kendall notes. However, Seneta (1983) points out that Abbe did write two short statistical articles, in 1878 and 1895. In his remarkable career in optics, initiated by Carl Zeiss, he was motivated to examine the properties of counts taken with the new Zeiss haemacytometer. His Poisson model for the distribution of counts was of immediate interest to users of the instrument. Although anticipating Student (1907), these articles were overlooked by statisticians.

All three of Abbe's statistical papers are reprinted in his collected works (Abbe, 1904–1906), where further biographical information is given. See also E. Seneta's 1982 biography of Abbe in the *Encyclopedia of Statistical Sciences*.

References

Abbe, E. (1863). Ueber die Gesetzmässigkeit in der Vertheilung der Fehler bei Beobachtungsreihen. Habilitationsschrift, Jena. [Reprinted in Abbe (1906), Vol. 2, pp. 55–81.]

Abbe, E. (1904–1906). *Gesammelte Abhundlungen*, Vols. 1 3. Gustav Fis cher, Jena.

Anderson, R.L. (1942). Distribution of the serial correlation coefficient. *Ann. Math. Statist.*, **13**, 1–13.

Bienaymé, I.-J. (1852). Sur la probabilité des erreurs d'après la méthode des moindres carrés. *J. Math. Pures Appl.*, **17**, 33–78.

David, H.A. (1998). Early measures of variability. *Statist. Science*, **13**, 368–377.

Gauss, C.F. (1809). *Theoria motus*. Eng. trans. by C.H. Davis (1857). Little, Brown Boston.

Hald, A. (1998). *A History of Mathematical Statistics from 1750 to 1930*. Wiley, New York.

Helmert, F.R. (1907). *Die Ausgleichungsrechnung nach der Methode der kleinsten Quadrate*. 2nd edn. Teubner, Leipzig.

Kendall, M.G. (1971). The work of Ernst Abbe. *Biometrika*, **58**, 369–373.

Lancaster, H.O. (1966). Forerunners of the Pearson χ^2. *Austral. J. Statist.*, **8**, 117–126.

Neumann, J. von, Kent, R.H., Bellinson, H.R., and Hart, B.I. (1941). The mean square successive difference. *Ann. Math. Statist.*, **12**, 153–162.

Seneta, E. (1983). Modern probabilistic concepts in the work of E. Abbe and A. de Moivre. *Math. Scientist.*, **8**, 75–80.

Student (1907). On the error of counting with a haemacytometer. *Biometrika*, **5**, 351–360.

On the Conformity-to-a-Law of the Distribution of Errors in a Series of Observations

E. Abbe

1

In order to specify completely the assumptions underlying the discussion to follow, we will suppose that we are dealing with some process in which the value of a measurable characteristic B varies with another, A (as, for example, the pressure of a body of steam varies with temperature, or the like). Then, on the one hand, for a series of values $a_1, a_2 \ldots a_n$ of the determining variable A, let the corresponding values $b_1, b_2 \ldots b_n$ of B be obtained by direct measurement; on the other hand, suppose that some theoretical formula, expressing the dependence as $B = F(A)$, gives calculated values $\beta_1, \beta_2, \ldots, \beta_n$ corresponding to $a_1, \ldots a_n$.[1] The question is now: what is the probability that the n differences

$$\beta_1 - b_1 - x_1 \ldots \beta_n \quad b_n - x_n$$

arise entirely from random errors of observation?

It is clear that the answer to this question must begin with a specific assumption on the relative frequency of random errors of various sizes for the observational process in question. We confine ourselves here to the use of a law that Gauss has derived from simple considerations and that has proved itself in many comparisons with extensive series of observations by always providing a close approximation to reality. According to this law, the probability w of the occurrence of an error whose size lies between x and $x + dx$ (dx being a differential in the usual sense) is determined by the formula

$$w = \frac{h}{\sqrt{\pi}} e^{-h^2 x^2} dx. \tag{1}$$

---

[1] Situations occurring in practice can in most cases be reduced to this simple scheme, even when more than two variables are involved.

The constant h, the measure of precision of the observational process, is linked with the probable error ρ of an observation by the equation

$$h\rho = 0.4769\ldots.$$

With these assumptions, the above question may be formulated as follows: if the precision of the observations is known through the corresponding value of h (or ρ), with what degree of probability can we expect the chance occurrence of a set of errors such as the given $x_1, x_2, \ldots x_n$? For, the probability of this event measures at the same time the probability with which the set of errors, after they have occurred, may be regarded as having arisen by chance.

What will the probability (i.e., the relative frequency) of a set of error variables depend on? It is clear that two different features of the set are here of primary importance, namely the absolute value of the individual errors and the pattern of successive errors. (Of course, we regard the observations as ordered according to the values of the argument that plays the role of the independent variable in the functional relationship.) A large error, relative to the value of h or ρ, occurs according to equation (1) more rarely than a small one. Consequently, a set consisting mainly of large differences can occur only relatively seldom. On the other hand, a set of errors whose members lie predominantly on one side or show a certain regularity of arrangement are to be expected with only a low probability.

We can now attempt to deal separately with these two considerations when forming our assessment. However, this can be carried out only with a certain arbitrariness, yet in such a way that, on the whole, the main features of the data are brought out.

For this purpose we will introduce two functions of the n values $x_1, x_2, \ldots x_n$. One of these is formed in such a way that its value depends only on the absolute values of the x's, whereas the second, on the contrary, is independent of the individual magnitudes of the x's, but varies with the differences existing among the x's, and with their ordering. Let us imagine two functions constructed according to these requirements. In view of the values these functions assume for any given set of x's, they will be suitable for distinguishing different sets from one another. They will do this by separating, at any rate, sets to which they immediately assign substantially different probabilities.

The conditions placed on the first function are evidently satisfied most simply by the sum of the squares of the n errors

$$\sum x^2 = x_1^2 + x_2^2 \ldots + x_n^2 = \Delta.$$

Use of this function for the intended purpose recommends itself all the more since Δ arises naturally in the application of the method of least squares. For construction of the other function, which is to characterize the way in which the quantity $\sum x^2$ is divided among its n terms, we are guided by the following simple considerations.

As happens always, a preponderance of errors on one side, with a noticeable trend, is viewed as indicating the presence of apparently regular causes of the deviations. Conversely, a natural indicator of a random origin should reflect the maximum possible irregularity in the relative size of the errors and in their ordering. But it is easy to set down an exact expression that specifies mathematically the degree of such irregularity. We see at once that, e.g., the sum of all differences between two successive errors, with all differences taken as positive in the calculation, decreases in magnitude the more the pattern of the n variables approaches some particular form of regularity, and vice versa; if, in order to remove signs arithmetically, we take the sum of squares of the differences, i.e., the expression

$$(x_1 - x_2)^2 + (x_2 - x_3)^2 \ldots + (x_{n-1} - x_n)^2,$$

the same applies to this also.

We now have a function of the n quantities x that (a) is independent of the actual magnitudes in that a simultaneous change in each by an arbitrary amount leaves its value unchanged, (b) attains a smaller (larger) value, the less (more) the n terms (apart from their order) differ from one another, and finally which (c) for given values of the x's goes from a minimum to a maximum as the x's, from being taken in order of magnitude, are rearranged so that two neighboring x's are as different as possible. The arrangement in (c) is to be such that every change in order, with the exception of a simple reversal, leads to a change in the value of the function (unless there exist special relations among the terms).

Finally, in order to simplify subsequent developments, we add an nth term to the above expression, namely the square of the difference between the last and the first x, so that the complete expression becomes

$$(x_1 - x_2)^2 + (x_2 - x_3)^2 \ldots + (x_{n-1} - x_n)^2 + (x_n - x_1)^2 = \Theta.$$

The statements just made apply here also, except for the unimportant modification that now any cyclic rearrangement of the n errors produces no change in value. Accordingly, the sum of squares of the differences Θ, formed in this way, takes into account all the circumstances that determine an immediate verdict on the probability of a set of errors judged by the manner of their distribution.

If Θ is used in combination with the sum Δ of the squares of the errors in order to distinguish between different sets of errors, we can restate the problem described above as follows: for an arbitrary number of observations, what is the probability a priori of the occurrence of a set of errors for which the function Δ attains a specified value (or lies between specified limits) and simultaneously the function Θ attains an also specified different value (or lies between specified limits)? With the solution of this problem there will be no further difficulties in assessing the probability of a random origin of a given set of errors.

2

The first step in dealing with the analytical task in question is to obtain the probability that for a sequence of n observations Δ falls between two specified limits, with Θ left unspecified.

Suppose the relative frequency of an error of observation depends on its magnitude according to equation (1). Then according to the laws of combinatorics the probability of the joint occurrence of n separately determined errors $x_1, \ldots x_n$, permitted to vary in turn by $dx_1 \ldots dx_n$, is

$$dx_1 dx_2 \ldots dx_n \frac{h^n}{\pi^{n/2}} \, e^{-h^2(x_1^2 + x_2^2 \ldots + x_n^2)}.$$

Consequently the probability of a set of errors for which the sum of the squares of the errors lies between the limits 0 and Δ^1 is given by

$$\Phi(\Delta) = \frac{h^n}{\pi^{n/2}} \int dx_1 \int dx_2 \ldots \int dx_n \, e^{-h^2(x_1^2 + x_2^2 \ldots + x_n^2)}, \qquad (2)$$

where the integration extends over all values of the n variables satisfying the condition

$$0 < x_1^2 + x_2^2 \ldots + x_n^2 < \Delta.$$

The preceding integral has already been evaluated in more than one way. Nevertheless we employ yet another way, using techniques which serve as preparation for what follows and which lead particularly easily and simply to our immediate aim.

We will make use of the well-known Dirichlet discontinuity factor method. However, instead of the integral employed by Dirichlet, we will use another integral that has significant advantages for present purposes. Its introduction never causes individual elements to become infinite and does not necessitate the addition of imaginary elements that must be eliminated in the end result.

Consider the following integral, defined over real values,

$$\int_{-\infty}^{\infty} \frac{e^{\sigma \phi i}}{a + \phi i} \, d\phi \quad (i = \sqrt{-1}).$$

With the aid of theorems established by Cauchy about integration over complex values of the variables, one can easily satisfy oneself that the integral is $2\pi e^{-\sigma a}$ for σ positive and zero for σ negative, provided a is a positive

---

[1]There is evidently no loss of generality in immediately taking zero as one of the limits.

constant, different from zero. For, the integral is single-valued for all complex ϕ and becomes infinite only for $\phi = ai$. Also, the integral vanishes when the integration extends from $-\infty$ to $+\infty$ through complex values with infinite modulus, whose imaginary component has the same sign as σ. This can be seen at once on setting

$$\phi = r \cos \psi + r \sin \psi i$$

and letting r tend to infinity. But since the integral taken over real values differs from the latter integral only by the residual-integrals for the poles lying between the two paths of integration, it follows easily that the original integral must vanish for σ negative and must equal $2\pi e^{-\sigma a}$ for σ positive. Dividing by this last value, we obtain therefore

$$\frac{1}{2\pi} \int_{-\infty}^{\infty} \frac{e^{\sigma(a+\phi i)}}{(a + \phi i)}\, d\phi = 1, \text{ for } \sigma > 0; = 0 \text{ for } \sigma < 0. \tag{3}$$

Hence we have an integral that can be used directly as discontinuity factor for the integration in (2)[1].

On setting

$$\sigma = \Delta - \left(x_1^2 + x_2^2 \dots + x_n^2\right)$$

and multiplying (2) by the above integral, each of the n variables can be integrated from $-\infty$ to ∞, since the added factor eliminates all regions for which the condition $x_1^2 + x_2^2 \dots + x_n^2 < \Delta$ is not satisfied. After reversing the order of integration with respect to ϕ, for which evidently all required conditions hold, and after separation of the individual variables, we obtain

$$\Phi(\Delta) = \frac{1}{2\pi} \frac{h^n}{\pi^{n/2}} \int_{-\infty}^{\infty} d\phi\, \frac{e^{\Delta(a+\phi i)}}{a + \phi i}$$
$$\cdot \int_{-\infty}^{\infty} dx_1\, e^{-(h^2+a+\phi i)x_1^2} \dots \int_{-\infty}^{\infty} dx_n\, e^{-(h^2+a+\phi i)x_n^2}.$$

But

$$\int_{-\infty}^{\infty} dx_k\, e^{-(h^2+a+\phi i)x_k^2} = \frac{\sqrt{\pi}}{\sqrt{h^2 + a + \phi i}}$$

so that

$$\Phi(\Delta) = \frac{h^n}{2\pi} \int_{-\infty}^{\infty} d\phi\, \frac{e^{\Delta(a+\phi i)}}{(a + \phi i)(h^2 + a + \phi i)^{n/2}}. \tag{4}$$

This integral can be further reduced by the substitution

$$\frac{1}{(h^2 + a + \phi i)^{n/2}} = \frac{1}{\Gamma(n/2)} \int_0^{\infty} dy\, e^{-(h^2+a+\phi i)y}\, y^{n/2-1},$$

---

[1] The use of an integral similar to the above as discontinuity factor is familiar to me through the lectures of Professor Riemann in Göttingen.

which gives after reversal of the order of integration

$$\Phi(\Delta) = \frac{h^n}{2\pi\Gamma(n/2)} \int_0^\infty dy\, e^{-h^2 y} y^{\frac{n}{2}-1} \int_{-\infty}^\infty d\phi\, \frac{e^{(\Delta-y)(a+\phi i)}}{a+\phi i}.$$

Here the integral w.r.t. ϕ, which appears as a multiplier of the preceding integral, is, in conjunction with $1/(2\pi)$, simply the original discontinuity factor (3) with $\sigma = (\Delta - y)$. It is therefore $= 1$, as long as $y < \Delta$, and $= 0$, if $y > \Delta$. Consequently the expression reduces finally to

$$\Phi(\Delta) = \frac{h^n}{\Gamma(n/2)} \int_0^\Delta e^{-h^2 y} y^{\frac{n}{2}-1} dy, \tag{5}$$

as can also be obtained by other methods.

By means of this formula we can immediately state the probability of a set of errors having a sum of squares between two arbitrary limits Δ_1 and Δ_2. There follows in particular the probability that this sum falls between Δ and $\Delta + d\Delta$:

$$\phi(\Delta) = \frac{h^n}{\Gamma(n/2)} e^{-h^2\Delta} \Delta^{\frac{n}{2}-1} d\Delta, \tag{6}$$

where $\phi(\Delta)$ has been used to distinguish this case. This equation now represents the relative frequency of a set of errors as a function of Δ similarly to the way eq. (1) represents the relative frequency of a single error as a function of its magnitude.

If in (6) we replace Δ by the so-called mean error upon setting

$$\Delta = n\delta^2; \quad d\Delta = 2n\delta d\delta,$$

then the expression for the probability of a set of errors with mean error between δ and $\delta + d\delta$ is

$$\phi(\delta) = \frac{2n^{n/2}h^n}{\Gamma(n/2)} e^{-nh^2\delta^2} \delta^{n-1} d\delta. \tag{7}$$

For large n, according to a well-known approximation for the Γ-function, namely

$$\Gamma(\mu) = \sqrt{\frac{2\pi}{\mu}} \left(\frac{\mu}{e}\right)^\mu,$$

(7) may be expressed approximately as

$$\phi(\delta) = \sqrt{\frac{n(2e)^n}{\pi}} h^n e^{-nh^2\delta^2} \delta^{n-1} d\delta. \tag{8}$$

3

We proceed now to determine the probability of a set of errors for which the sum of the squares of the successive differences, i.e., the function Θ, falls between given limits, with Δ remaining unspecified.

After some further preparation this problem can be treated in a similar manner to the above with the help of the foregoing techniques of integration. If, as above, zero is taken as one of the limits and an arbitrary value Θ as the other, the expression for the required probability is

$$\Psi(\Theta) = \frac{h^n}{\pi^{n/2}} \int dx_1 \ldots \int dx_n \, e^{-h^2(x_1^2 + x_2^2 \ldots + x_n^2)},$$

where the integration now extends over all values of the variables $x_1 \ldots x_n$, that satisfy the condition

$$(x_1 - x_2)^2 + (x_2 - x_3)^2 \ldots + (x_{n-1} - x_n)^2 + (x_n - x_1)^2 < \Theta.$$

If we multiply the integrand by the discontinuity factor of eq. (3), with σ taken as

$$\sigma = \Theta - [(x_1 - x_2)^2 \ldots + (x_n - x_1)^2],$$

we again obtain an $(n + 1)$-fold integral with constant limits $-\infty$ and ∞. After reversal of the order of integration this yields the expression

$$\Psi(\Theta) = \frac{1}{2\pi} \frac{h^n}{\pi^{n/2}} \int_{-\infty}^{\infty} \frac{d\phi}{a + \phi i} \, e^{\Theta(a + \phi i)}$$

$$\cdot \int_{-\infty}^{\infty} dx_1 \ldots \int_{-\infty}^{\infty} dx_n \, e^{-h^2(x_1^2 + \ldots + x_n^2) - (a + \phi i)[(x_1 - x_2)^2 \ldots + (x_n - x_1)^2]}.$$

$$(9)$$

Here the variables $x_1 \ldots x_n$ can no longer be immediately separated, for

$$(x_1 - x_2)^2 \ldots + (x_n - x_1)^2$$

$$= 2(x_1^2 + x_2^2 \ldots + x_n^2) - 2(x_1 x_2 + x_2 x_3 \ldots + x_{n-1} x_n + x_n x_1)$$

$$(10)$$

and in the second part of this expression, which occurs in the exponent of the above integral, the x's are interlinked. But the separation can be effected by a transformation of the integral through the introduction of new variables.

It is a well-known result in the theory of homogeneous quadratic functions—which evidently include the expression in (10)—that any such function of n variables $x_1 \ldots x_n$ can be changed by a linear transformation

$$x_k = \beta_{k,1} \xi_1 + \beta_{k,2} \xi_2 \ldots + \beta_{k,n} \xi_n \tag{11}$$

into another function of the form

$$\alpha_1 \xi_1^2 + \alpha_2 \xi_2^2 \ldots + \alpha_n \xi_n^2.$$

This can always be done by a so-called orthogonal transformation, so that at the same time one has

$$x_1^2 + x_2^2 \ldots + x_n^2 = \xi_1^2 + \xi_2^2 \ldots + \xi_n^2.$$

To effect the separation of the variables in the above integral, it remains only to introduce n new variables $\xi_1, \ldots \xi_n$ that are linked to the x's by n equations of the form (11), and to determine the n^2 coefficients $\beta_{1,1} \ldots \beta_{n,n}$ such that the above conditions are satisfied. Then the following transformations result,

$$x_1^2 \cdots + x_n^2 \quad \text{into} \quad \xi_1^2 \cdots + \xi_n^2$$

$$(x_1 - x_2)^2 \cdots + (x_n - x_1)^2 \quad \text{into} \quad \alpha_1 \xi_1^2 + \alpha_2 \xi_2^2 \cdots + \alpha_n \xi_n^2$$

so that the integrand in (9) assumes the form

$$e^{-h^2 (\xi_1^2 + \xi_2^2 \cdots + \xi_n^2) - (a+\phi i)(\alpha_1 \xi_1^2 + \alpha_2 \xi_2^2 \cdots + \alpha_n \xi_n^2)}$$

$$= e^{-[h^2 + \alpha_1 (a+\phi i)] \xi_1^2} \cdot e^{-[h^2 + \alpha_2 (a+\phi i)] \xi_2^2} \cdots e^{-[h^2 + \alpha_n (a+\phi i)] \xi_n^2}.$$

The coefficients $\alpha_1 \ldots \alpha_n$ are still to be determined.

When introducing new variables in a multiple integral we must also, following familiar rules, multiply the integrand by the functional determinant of the n functions that replace the original variables. Since these functions are linear here, the functional determinant reduces to the determinant of the set of coefficients $\beta_{1,1} \cdots \beta_{n,n}$ and is consequently identically $= 1$, if the linear transformation is prescribed to be orthogonal. The integrand is otherwise unchanged; $d\xi_1 d\xi_2 \cdots d\xi_n$ simply replaces $dx_1 dx_2 \cdots dx_n$. Finally, the limits of integration also remain the same. For, since reciprocally, each of the new variables can be represented as a linear function of the original variables, each new variable must range over the entire interval from $-\infty$ to $+\infty$ if all the original variables do so.

The expression for $\Psi(\Theta)$ accordingly takes the form

$$\Psi(\Theta) = \frac{1}{2\pi} \frac{h^n}{\pi^{n/2}} \int_{-\infty}^{\infty} \frac{d\phi}{a + \phi i} e^{\Theta(a+\phi i)}$$

$$\cdot \int_{-\infty}^{\infty} d\xi_1 e^{-[h^2 + \alpha_1 (a+\phi i)] \xi_1^2} \cdots \int_{-\infty}^{\infty} d\xi_n e^{-[h^2 + \alpha_n (a+\phi i)] \xi_n^2}.$$

The integrations for $\xi_1 \ldots \xi_n$ can now be performed by means of the equation already used earlier

$$\int_{-\infty}^{\infty} e^{-Hx^2} dx = \frac{\sqrt{\pi}}{\sqrt{H}}.$$

Its validity requires only that the real part of H be positive. This condition is evidently satisfied even if the coefficients $\alpha_1 \ldots \alpha_n$ should receive negative values, since in that case the completely arbitrary constant a (which must simply be taken as nonzero, so that the function does not become infinite upon integration) can be chosen small enough to ensure that always

$$h^2 + \alpha_k\, a > 0.$$

We then obtain

$$\Psi(\Theta) = \frac{h^n}{2\pi}\int_{-\infty}^{\infty} d\phi \frac{e^{\Theta(a+\phi i)}}{(a+\phi i)\sqrt{(h^2+\alpha_1(a+\phi i))\ldots(h^2+\alpha_n(a+\phi i))}}. \quad (12)$$

To continue the evaluation of this integral, we first determine the values of the n coefficients $\alpha_1 \ldots \alpha_n$. This is easily done with the help of the following theorems proved by Cauchy, Jacobi, and others. Suppose a homogeneous function of the second degree in the n variables $x_1 \ldots x_n$, of general form

$$\sum^{k,m} b_{k,m} x_k x_m \quad \text{with} \quad b_{k,m} = b_{m,k},$$

is transformed by an orthogonal transformation into

$$\alpha_1 \xi_1^2 + \alpha_2 \xi_2^2 \cdots + \alpha_n \xi_n^2.$$

Then the coefficients $\alpha_1 \ldots \alpha_n$ are the roots, always all real, of the equation of the nth degree

$$F(\zeta) = 0,$$

where $F(\zeta)$ represents the determinant[1]

$$\begin{vmatrix} b_{1,1} - \zeta & b_{1,2} & \cdots & b_{1,n} \\ b_{2,1} & b_{2,2} - \zeta & \cdots & b_{2,n} \\ & & & \\ b_{n,1} & b_{n,2} & \cdots & b_{n,n} - \zeta \end{vmatrix}.$$

In the present case the transformed function is

$$2\left(x_1^2 + x_2^2 \cdots + x_n^2\right) - 2\left(x_1 x_2 + x_2 x_3 \cdots + x_{n-1} x_n + x_n x_1\right)$$

so that the coefficients from which the determinant is to be constructed are:

$$b_{1,1} = b_{2,2} \cdots = b_{n,n} = 2$$

$$b_{1,2} = b_{2,1} = -1, \; b_{2,3} = b_{3,2} = -1; \; \ldots b_{n-1,n} = b_{n,n-1} = -1,$$

$$b_{n,1} = b_{1,n} = -1$$

---

[1] J. Baltzer. *Theorie der Determ.* (*Theory of Determinants*, presented with reference to the original sources) (1857). Leipzig: S. Hirzel.

with all other coefficients $= 0$. Hence the constants $\alpha_1 \ldots \alpha_n$ appear as the roots of the equation

$$0 = \begin{vmatrix} 2-\zeta & -1 & 0 & & -1 \\ -1 & 2-\zeta & -1 & & 0 \\ 0 & -1 & 2-\zeta & & 0 \\ \cdots & \cdots & \cdots & \cdots & \cdots \\ -1 & 0 & 0 & & 2-\zeta \end{vmatrix}.$$

By means of this relation the roots are not difficult to obtain. The form of the determinant is evidently characterized by the fact that each row and each column can be derived from the first by cyclic development of the n terms

$$2-\zeta \quad -1 \quad 0 \quad \cdots \quad 0 \quad -1.$$

But a determinant with this property can, according to a theorem of Spottiswoode, always be represented as a product of n factors. If

$$k_1 \quad k_2 \quad \cdots \quad k_n$$

are the elements of the first row, the product is

$$
\begin{aligned}
& k_1 + \mu k_2 + \mu^2 k_3 \cdots + \mu^{n-1} k_n \\
\times\ & (k_1 + \mu^2 k_2 + \mu^4 k_3 \cdots + \mu^{2(n-1)} k_n) \\
\times\ & (k_1 + \mu^3 k_2 + \mu^6 k_3 \cdots + \mu^{3(n-1)} k_n) \\
& \cdots\cdots\cdots\cdots\cdots\cdots\cdots\cdots \\
\times\ & (k_1 + k_2 + k_3 \cdots + k_n),
\end{aligned}
$$

where μ represents an arbitrary complex root of the equation $\mu^n = 1$.[1] In the present case, with

$$k_1 = 2 - \zeta;\ k_2 = -1;\ k_3 = k_4 \cdots = k_{n-1} = 0;\ k_n = -1,$$

this product becomes

$$(2 - (\mu + \mu^{n-1}) - \zeta)(2 - (\mu^2 + \mu^{2(n-1)}) - \zeta) \cdots (2 - (\mu^n + \mu^{n(n-1)}) - \zeta).$$

Then the n roots of the equation in ζ, which results when the product is

---

[1] A proof of this theorem can be found in *Zeitschr. f. Math. u. Phys.*, Vol. VII, 1862, p. 440. [G. Zehfuss: Applications of a special determinant, idem, pp. 439–445.]

set $= 0$; i.e., the coefficients $\alpha_1 \ldots \alpha_n$ are of the form

$$\alpha_k = 2 - (\mu^k + \mu^{k(n-1)}) = 2 - (\mu^k + \mu^{-k}).$$

Finally, if μ, according to its meaning, is replaced by one of the roots of unity, e.g.,

$$\mu = \cos \frac{2\pi}{n} + i \sin \frac{2\pi}{n},$$

we have

$$\mu^k + \mu^{-k} = 2 \cos k \frac{2\pi}{n}$$

and hence

$$\alpha_k = 2(1 - \cos k \frac{2\pi}{n}) = 4 \sin^2 k \frac{\pi}{n}. \tag{13}$$

Because of the relations existing among the n coefficients $\alpha_1 \ldots \alpha_n$ as a result of this formula, the integral in (12) can either be evaluated completely or at least reduced to a substantially simpler form. For each n we have

$$\alpha_n = 0; \quad \alpha_{n-1} = \alpha_1; \quad \alpha_{n-2} = \alpha_2; \quad \ldots.$$

Consequently one of the factors under the radical sign in eq. (12) always reduces to h^2. The remaining factors are either equal in pairs, if n is odd $= 2\nu + 1$, or they are equal in pairs, with the exception of one, if n is even $= 2\nu + 2$. In the latter case the term with

$$\alpha_n = 2(1 - \cos \pi) = 4$$

occurs only once. Accordingly we obtain
  1) for $n = 2\nu + 1$

$$\Psi(\Theta) = \frac{h^n}{2\pi} \frac{1}{} \int_{-\infty}^{\infty} d\phi \frac{e^{\Theta(a+\phi i)}}{(a + \phi i)(h^2 + \alpha_1(a + \phi i)) \ldots (h^2 + \alpha_\nu(a + \phi i))}$$

$$= \frac{h^{n-1}}{2\pi . \alpha_1 . \alpha_2 \ldots \alpha_\nu} \int_{-\infty}^{\infty} d\phi \frac{e^{\Theta(a+\phi i)}}{(a + \phi i)(\frac{h^2}{\alpha_1} + a + \phi i) \ldots (\frac{h^2}{\alpha_\nu} + a + \phi i)};$$
$$\tag{14a}$$

  2) for $n = 2\nu + 2$

$$\Psi(\Theta) = \frac{h^{n-1}}{2\pi \cdot \alpha_1 \alpha_2 \ldots \alpha_\nu}$$

$$\int_{-\infty}^{\infty} d\phi \frac{e^{\Theta(a+\phi i)}}{(a + \phi i)(\frac{h^2}{\alpha_1} + a + \phi i) \ldots (\frac{h^2}{\alpha_\nu} + a + \phi i)\sqrt{h^2 + 4(a + \phi i)}}.$$
$$\tag{14b}$$

In both cases

$$\alpha_1 = 2\left(1 - \cos\frac{2\pi}{n}\right); \ \ldots \ \alpha_\nu = 2\left(1 - \cos\nu\frac{2\pi}{n}\right),$$

but $\nu = \frac{n-1}{2}$ if n is odd, and $\nu = \frac{n-2}{2}$ if n is even.

In the first case the integral can be evaluated completely with the help of the theorems already used above concerning the integration of single-valued functions over complex values of the variables. The integrand in (14a) is single-valued and always remains finite except for those values of ϕ for which $(a + \phi i) = 0$, $-\frac{h^2}{\alpha_1}$, $\frac{-h^2}{\alpha_2} \ldots \frac{-h^2}{\alpha_\nu}$ or $\phi = ai$, $\left(a + \frac{h^2}{\alpha_1}\right)i \ldots \left(a + \frac{h^2}{\alpha_\nu}\right)i$. Moreover, we see that the integrand vanishes for complex values of ϕ having infinite modulus with positive imaginary component; for, if we put

$$\phi = r\cos\psi + r\sin\psi i,$$

then

$$e^{\Theta(a+\phi i)} = e^{\Theta a} \cdot e^{\Theta r\cos\psi i} \cdot e^{-\Theta r\sin\psi}.$$

As long as $\sin\psi$ or the imaginary part of ϕ is positive, this product becomes vanishingly small with increasing r, since Θ always remains positive.

Accordingly, the integral of the function occurring in (14a), taken from $-\infty$ to ∞ through complex values of ϕ with infinite modulus and positive imaginary component, must vanish. But this last integral has beginning and endpoint of integration in common with the integral through real values that constitutes the expression for $\Psi(\Theta)$. The difference (the former subtracted from the latter) must therefore be equal to the sum of the integrals over all the poles lying between the paths of integration, which is equal to $2\pi i$ times the sum of the residuals at the various poles. Since the quantity to be subtracted in this difference vanishes in view of the foregoing, it follows that the integral appearing in (14a) must itself have this sum as its value.

The poles lying between the paths of integration consist of all values of ϕ for which one of the factors in the denominator is zero, since without exception these values are on the positive imaginary axis, namely

$$\phi = ai; \ \phi = \left(a + \frac{h^2}{\alpha_1}\right)i; \ \ldots \ \phi = \left(a + \frac{h^2}{\alpha_\nu}\right)i.$$

Through the application of familiar arguments the integral corresponding to any of these values is given by $2\pi.R$, where R denotes the product of all factors in the function that remain finite for the value in question. For the first pole, $\phi = ai$, we have accordingly

$$2\pi.R = \frac{2\pi}{\frac{h^2}{\alpha_1} \cdot \frac{h^2}{\alpha_2} \ldots \frac{h^2}{\alpha_\nu}} = 2\pi\frac{\alpha_1 \cdot \alpha_2 \cdots \alpha_\nu}{h^{n-1}}.$$

For any of the ν other poles, for which $\phi = \left(\frac{h^2}{\alpha_k} + a\right)i$, say, we have

$$2\pi . R_k = \frac{2\pi . e^{-\Theta\frac{h^2}{\alpha_k}}}{-\frac{h^2}{\alpha_k}\left(\frac{h^2}{\alpha_1} - \frac{h^2}{\alpha_k}\right)\cdots\left(\frac{h^2}{\alpha_{k-1}} - \frac{h^2}{\alpha_k}\right)\left(\frac{h^2}{\alpha_{k+1}} - \frac{h^2}{\alpha_k}\right)\cdots\left(\frac{h^2}{\alpha_\nu} - \frac{h^2}{\alpha_k}\right)}$$

$$= -2\pi . \frac{\alpha_1\cdot\alpha_2\cdots\alpha_\nu\alpha_k^{\nu-1}e^{-\Theta\frac{h^2}{\alpha_k}}}{h^{n-1}(\alpha_k - \alpha_1)\cdots(\alpha_k - a_{k-1})(\alpha_k - \alpha_{k+1})\cdots(\alpha_k - \alpha_\nu)}.$$

Taking into account the constant factors occurring in (14a) we obtain now

$$\Psi(\Theta) = 1 - \sum_{1,\nu}^{k} \frac{\alpha_k^{\nu-1}e^{-\Theta\frac{h^2}{\alpha_k}}}{(\alpha_k - \alpha_1)\cdots(\alpha_k - \alpha_\nu)}, \tag{15a}$$

where the summation is to be taken over all k from 1 to $\nu = \frac{n-1}{2}$.

We turn now to the other case, n even $= 2\nu + 2$. Here also the expression for $\Psi(\Theta)$ may be reduced with the aid of similar considerations if the irrational factor in the integral is first removed. This can easily be accomplished by representing this factor in the form of a definite integral, namely

$$\frac{1}{2\sqrt{\frac{h^2}{4} + a + \phi i}} = \frac{1}{2\sqrt{\pi}}\int_0^\infty e^{-(\frac{h^2}{4}+a+\phi i)u}\frac{du}{\sqrt{u}}.$$

Reversal of the order of integration, which clearly needs no further justification here, then yields:

$$\Psi(\Theta) = \frac{h}{2\sqrt{\pi}}\int_0^\infty e^{-\frac{h^2}{4}u}\frac{du}{\sqrt{u}}$$

$$\cdot\left\{\frac{h^{n-2}}{2\pi\alpha_1\ldots\alpha_\nu}\int_{-\infty}^\infty d\phi\,\frac{e^{(\Theta-u)(a+\phi i)}}{(a+\phi i)\left(\frac{h^2}{\alpha_1} + a + \phi i\right)\cdots\left(\frac{h^2}{\alpha_\nu} + a + \phi i\right)}\right\}.$$

The expression in braces is now (allowing for the fact that here $\nu = \frac{n-2}{2}$) of the very form as that for $\Psi(\Theta)$ itself in the case treated first, except that $\Theta - u$ takes the place of Θ. As long as $\Theta - u$ is positive, i.e., $u < \Theta$, our previous arguments apply here without change and the expression can then be replaced by a sum of the form (15a). But if $u > \Theta$, so that $(\Theta-u)$ attains a negative value, a simple argument reveals that the integral with respect to ϕ must vanish. For if, in the exponential term, ϕi is multiplied by a negative factor, the exponential (and hence the entire integrand) becomes zero for complex values of ϕ with infinite modulus whose imaginary part is negative. Accordingly an integral extending from $-\infty$ to ∞ through such complex values evidently is zero. But the integral appearing in the above expression must also be zero, because the terminals of integration are the same and there are no poles between the paths of integration, such poles corresponding exclusively to positive imaginary values of ϕ.

Consequently, the expression for $\Psi(\Theta)$, for $n = 2\nu + 2$, reduces finally to the integral

$$\Psi(\Theta) = \frac{h}{2 \cdot \sqrt{\pi}} \int_0^\Theta e^{-(\frac{h}{2})^2 u} \frac{du}{\sqrt{u}} \left\{ 1 - \sum_{1,\nu}^k \frac{\alpha_k^{\nu-1} e^{-(\Theta-u)\frac{h^2}{\alpha_k}}}{(\alpha_k - \alpha_1) \cdots (\alpha_k - \alpha_\nu)} \right\}.$$

(15b)

The sum appearing in these equations can be represented in several other ways. For example, we may remark that this sum can quite generally be expressed as ratios of two determinants. For if we suppose all terms to be reduced to a common denominator, this will become the product of all differences that arise if in the sequence of ν values $\alpha_1, \ldots, \alpha_\nu$ we subtract from each α all subsequent ones. By a well-known theorem such a product is identical with the determinant

$$Q = \begin{vmatrix} 1 & \alpha_1 & \alpha_1^2 & \cdots & \alpha_1^{\nu-2} & \alpha_1^{\nu-1} \\ 1 & \alpha_2 & \alpha_2^2 & \cdots & \alpha_2^{\nu-2} & \alpha_2^{\nu-1} \\ \cdots & \cdots & \cdots & \cdots & \cdots & \cdots \\ 1 & \alpha_\nu & \alpha_\nu^2 & \cdots & \alpha_\nu^{\nu-2} & \alpha_\nu^{\nu-1} \end{vmatrix}.$$

The factors which accompany the individual numerators are now simply the coefficients in the expansion of Q by the elements in the last column. It follows directly that the entire numerator coincides with the determinant

$$P = \begin{vmatrix} 1 & \alpha_1 & \alpha_1^2 & \cdots & \alpha_1^{\nu-2} & f(\alpha_1) \\ 1 & \alpha_2 & \alpha_2^2 & \cdots & \alpha_2^{\nu-2} & f(\alpha_2) \\ \cdots & \cdots & \cdots & \cdots & \cdots & \cdots \\ 1 & \alpha_\nu & \alpha_\nu^2 & \cdots & \alpha_\nu^{\nu-2} & f(\alpha_\nu) \end{vmatrix}$$

if $f(\alpha_k)$ designates $\alpha_k^{\nu-1} e^{-\frac{h^2}{\alpha_k}\Theta}$ or $\alpha_k^{\nu-1} e^{-\frac{h^2}{\alpha_k}(\Theta-u)}$, respectively. Then $\sum = P/Q$, from which follows a result, for every set of coefficients $\alpha_1 \ldots \alpha_\nu$, that is required for the present special set in view of the meaning of $\Psi(\Theta)$: the sums become 1 if $f(\alpha_k) = \alpha_k^{\nu-1}$, i.e., if Θ or $(\Theta - u)$ is zero, respectively.

$$\alpha_k = 2\left(1 - \cos k\frac{2\pi}{n}\right) = 4\sin^2 k\frac{\pi}{n} \, ;$$

$$\alpha_k^{\nu-1} = 2^{2\nu-2} \cdot \sin^{2\nu-2} k\frac{\pi}{n};$$

$$(\alpha_k - \alpha_1)\cdots(\alpha_k - \alpha_\nu) = (-1)^{\nu-1}\cdot 2^{\nu-1}\left(\cos k\frac{2\pi}{n} - \cos\frac{2\pi}{n}\right)$$

$$\cdot\left(\cos k\frac{2\pi}{n} - \cos 2\frac{2\pi}{n}\right)\cdots\left(\cos k\frac{2\pi}{n} - \cos \nu\frac{2\pi}{n}\right).$$

By writing the trigonometric product in the form

$$\lim_{x=k\frac{\pi}{n}}\frac{\left(\cos 2x - \cos\frac{2\pi}{n}\right)\cdots\left(\cos 2x - \cos k\frac{2\pi}{n}\right)\cdots\left(\cos 2x - \cos \nu\frac{2\pi}{n}\right)}{\left(\cos 2x - \cos k\frac{2\pi}{n}\right)}$$

we find from a well-known formula, that[1]

$$(\alpha_k - \alpha_1)\cdots(\alpha_k - \alpha_\nu)$$

$$= (-1)^{\nu+k}\frac{n}{2^3\cdot\sin^2 k\frac{\pi}{n}\cdot\cos k\frac{\pi}{n}}, \quad \text{if } n = 2\nu + 1$$

$$= (-1)^{\nu+k}\frac{n}{2^4\cdot\sin^2 k\frac{\pi}{n}\cdot\cos^2 k\frac{\pi}{n}}, \quad \text{if } n = 2\nu + 2.$$

Finally, it follows that

$$n = 2\nu + 1: \Psi(\Theta) = 1 - \frac{2^n}{n}\sum_1^\nu(-1)^{\nu+k}e^{-\frac{h^2\Theta}{4\sin^2 k\frac{\pi}{n}}}\sin^{n-1}k\frac{\pi}{n}\cos k\frac{\pi}{n} \quad (16a)$$

$$n = 2\nu + 2: \Psi(\Theta) = \frac{h}{2\sqrt{\pi}}\int_0^\Theta e^{-\frac{h^2}{4}u}\frac{du}{\sqrt{u}}$$

$$\cdot\left\{1 - \frac{2^n}{n}\sum_1^\nu(-1)^{\nu+k}e^{-\frac{h^2(\Theta-u)}{4\sin^2 k\frac{\pi}{n}}}\sin^{n-2}k\frac{\pi}{n}\cos\frac{2k\pi}{n}\right\}. \quad (16b)$$

By differentiation with respect to Θ these formulae yield expressions that represent the relative frequency of different sets of errors as a function of Θ, the sum of squares of successive differences corresponding to the error sets.

4

Following the two problems treated so far, we still wish to investigate the nature of the dependence of the relative frequency of an error set on the relation between the values of the two functions Δ and Θ. For this purpose

---

[1]M.A. Stern, Lehrb. d. algebr. Analysis [Leipzig, C.F. Winter, 1860] p. 385.

we need an expression for the probability of occurrence of an error set for which

$$\Theta \le \mu \Delta.$$

Thus the n-fold integral

$$\frac{h^n}{\pi^{n/2}} \int dx_1 \ldots \int dx_n \ e^{-h^2(x_1^2 + \ldots + x_n^2)}$$

is to be taken over all those values of the n variables that satisfy the above inequality.

Referring to the transformation of the function Θ, carried out in the previous section by the introduction of the new set of variables $\xi_1 \ldots \xi_n$, we can immediately write the above condition as

$$\mu \left(\xi_1^2 \cdots + \xi_n^2\right) - \left(\alpha_1 \xi_1^2 \cdots + \alpha_n \xi_n^2\right) = (\mu - \alpha_1)\xi_1^2 \cdots + (\mu - \alpha_n)\xi_n^2 > 0.$$

Also the integrals can be taken to lie between the constant limits $-\infty$ and ∞ if we introduce the integral (3) as discontinuity factor, with

$$\sigma = (\mu - \alpha_1)\xi_1^2 \cdots + (\mu - \alpha_n)\xi_n^2.$$

Accordingly, the expression for the probability, denoted by $X(\mu)$, of an error set satisfying the condition, becomes

$$X(\mu) \quad = \quad \frac{1}{2\pi} \frac{h^n}{\pi^{n/2}} \int_{-\infty}^{\infty} \frac{d\phi}{a + \phi i} \int_{-\infty}^{\infty} d\xi_1 \ e^{-(h^2 - (\mu - \alpha_1)(a + \phi i))\xi_1^2}$$
$$\cdots \int_{-\infty}^{\infty} d\xi_n e^{-(h^2 - (\mu - \alpha_n)(a + \phi i))\xi_n^2}.$$

Since the constant a, even though assumed strictly positive, can always be specified so that $(h^2 - (\mu - \alpha_k)a)$ remains positive, the integrations over the variables $\xi_1 \ldots \xi_n$ can be performed by means of the formula used earlier, giving

$$X(\mu) \quad = \quad \frac{h^n}{2\pi}$$
$$\cdot \int_{-\infty}^{\infty} \frac{d\phi}{(a + \phi i)\sqrt{(h^2 - (\mu - \alpha_1)(a + \phi i)) \ldots (h^2 - (\mu - \alpha_n)(a + \phi i))}}.$$
$$\tag{17}$$

Note that $\alpha_n = 0$ and $\alpha_k = \alpha_{n-k}$, so that $(\mu - \alpha_n)$ reduces to μ and the other factors under the radical sign occur in pairs, either entirely or except for one factor, according as $n = 2\nu + 1$ or $2\nu + 2$. In the latter case the coefficient appearing only once is $\alpha_{\nu+1} = 4$. The above equation can therefore be written as

$$X(\mu) = \frac{h^n}{2\pi} \int_{-\infty}^{\infty}$$
$$\cdot \frac{d\phi}{\sqrt{h^2 - \mu(a+\phi i)}(a+\phi i)(h^2 + (\alpha_1 - \mu)(a+\phi i))\ldots(h^2 + (a_\nu - \mu)(a+\phi i))\sqrt{h^2 + (4-\mu)(a+\phi i)}},$$

where the last factor in the denominator, $\sqrt{h^2 + (4 - \mu)(a + \phi i)}$, is omitted if $n = 2\nu + 1$, but is retained if $n = 2\nu + 2$.

The above integral may be evaluated by the method used for eq. (14). The irrational factors are again replaced by the corresponding Eulerian integrals, resulting in a single-valued function after interchange of the order of integration. For the first case, $n = 2\nu + 1$, we obtain in this way, after separation of variables,

$$X(\mu) = \frac{1}{\sqrt{\pi}} \int_0^\infty e^{-h^2 u} \frac{du}{\sqrt{u}} \frac{h^n}{2\pi}$$
$$\cdot \int_{-\infty}^\infty \frac{e^{\mu u(a+\phi i)} d\phi}{(a + \phi i)(h^2 + (\alpha_1 - \mu)(a + \phi i))\ldots(h^2 + (\alpha_\nu - \mu)(a + \phi i))}.$$

Since (μu) in this expression is essentially positive, the arguments that led to the evaluation of $\Psi(\Theta)$ in eq. (14a) can be applied without change to the integration with respect to ϕ. Accordingly, the latter integral reduces to $2\pi i$ times the sum of the residues of the integrand at all poles that correspond to a value of ϕ with positive imaginary part. But the values of the variable for which the integrand becomes infinite are $\phi = ai$, $\phi = (\frac{h^2}{\alpha_1 - \mu} + a)i$; $\ldots \phi = (\frac{h^2}{\alpha_\nu - \mu} + a)i$. Now the constant a must be chosen so that $(h^2 + (\alpha_k - \mu)a)$ remains positive for all k, giving $(\frac{h^2}{\alpha_k - \mu} + a)$ the same sign as $\alpha_k - \mu$. Thus, the above values of ϕ having a positive imaginary part are those for which $\alpha_k > \mu$.

The total sum of residue integrals in the present case can now be easily derived from expression (15a) which represents such a sum for the very similar integral in (14a). We need only replace α by $(\alpha - \mu)$ and Θ in the exponent by μu, as well as add the factor $\frac{1}{h^{n-1}}$ which disappears in the earlier case. This gives us

$$\frac{1}{h^{n-1}} \left[1 - \sum^k \frac{(\alpha_k - \mu)^{\nu-1} e^{-\mu u \frac{h^2}{\alpha_k - \mu}}}{(\alpha_k - \alpha_1)\ldots(\alpha_k - \alpha_\nu)} \right],$$

where all terms for which $\alpha_k < \mu$ must be omitted from the sum.

If the foregoing expression (together with the factor $\frac{1}{2\pi}$) is inserted into the above integral, the further integration over u can be performed term by term. It follows that

$$X(\mu) = \frac{h}{\sqrt{\pi}} \int_0^\infty e^{-h^2 u} \frac{du}{\sqrt{u}} - \sum^k \frac{(\alpha_k - \mu)^{\nu-1}}{(\alpha_k - \alpha_1)\dots(\alpha_k - \alpha_\nu)}$$

$$\cdot \frac{h}{\sqrt{\pi}} \int_0^\infty e^{-h^2(1+\frac{\mu u}{\alpha_k - \mu})} \frac{du}{\sqrt{u}}$$

$$= 1 - \sum^k \frac{(\alpha_k - \mu)^{\nu-1}\sqrt{1 - \frac{\mu}{\alpha_k}}}{(\alpha_k - \alpha_1)\dots(\alpha_k - \alpha_\nu)}, \tag{18a}$$

where the summation is to be restricted to those terms for which $\alpha_k > \mu$; i.e., $X(\mu)$ is the real part of the expression on the right.

From the original definition of the function Θ, as given in Section 1, we see directly that its value can never exceed four times the corresponding sum of squares Δ. More precisely, the transformed expression developed in Section 3 determines the upper limit of Θ as 4Δ for n even, but as $2(1+\cos\frac{\pi}{n})\Delta$ for n odd. Accordingly, the above expression for $X(\mu)$ reduces to unity through the vanishing of each term in the sum, as soon as $\mu \geq 2(1 + \cos\frac{\pi}{n})$. On the other hand, as a consequence of an earlier remark, $X(0) = 0$.

Next, we replace the coefficients $\alpha_1 \dots \alpha_\nu$ by their specific values, at the same time expressing μ in parallel form as $2(1 - \cos\kappa\frac{2\pi}{n})$, where κ varies continuously from 0 to ν. Then the probability that, for n odd, $\Theta \leqq \mu\Delta$, i.e., the ratio $\frac{\Theta}{\Delta} \leqq \mu$, is given by

$$X(\mu) = 1 - \frac{2^n}{n} \sum^k (-1)^{\nu+k} \sin\kappa\frac{\pi}{n}\cos\kappa\frac{\pi}{n}$$

$$\cdot \left(\sin(k+\kappa)\frac{\pi}{n}\sin(k-\kappa)\frac{\pi}{n}\right)^{\nu-1/2}, \tag{19a}$$

where the sum is again restricted to the terms that remain real, i.e., to those for which $k > \kappa$.

For the case $n = 2\nu + 2$ we can proceed similarly to the treatment of the corresponding case in Section 3, once the two irrational factors now appearing in (17) are replaced by definite integrals. Then $X(\mu)$ is given by two sums, formed analogously to (18a), whose real parts must still be integrated as in (15b). We will not pursue this further here.

[Translator: At this point we omit Abbe's lengthy argument, unnecessary for the modern reader, showing that the distribution of Θ/Δ is scale-independent. Abbe notes that (17) may consequently be simplified to

$$
X(\mu) =
$$
$$
\frac{1}{2\pi} \int_{-\infty}^{\infty} \frac{d\phi}{(a + \phi i)\sqrt{(1 - (\mu - \alpha_1)(a + \phi i)) \dots (1 - (\mu - \alpha_n)(a + \phi i))}} .]
$$
$$
(20)
$$

The behavior of $X(\mu)$ as a function of μ can be seen in broad terms by a simple argument. By differentiation of (20) with respect to μ we obtain the function that represents the relative frequency of a set of errors as a function of Θ/Δ. Denoted by $\chi(\mu)$, this function is then

$$
\chi(\mu) = d\frac{1}{4\pi} \int_{-\infty}^{\infty} \frac{d\phi}{\sqrt{(1 - (\mu - \alpha_1)(a + \phi i)) \dots (1 - (\mu - \alpha_n)(a + \phi i))}}
$$
$$
\cdot \left\{ \frac{1}{1 - (\mu - \alpha_1)(a + \phi i)} \dots + \frac{1}{1 - (\mu - \alpha_n)(a + \phi i)} \right\} .
$$

The foregoing integral does not depend on the constant a as long as a remains positive. With the disappearance of the factor $(a + \phi i)$, all terms in the denominator now remain finite, even for a vanishingly small. It becomes clear that we may set $a = 0$ to obtain the simpler expression

$$
\chi(\mu) = d\frac{1}{4\pi} \int_{-\infty}^{\infty} \frac{d\phi}{\sqrt{(1 - (\mu - \alpha_1)\phi i) \dots (1 - (\mu - \alpha_n)\phi i)}}
$$
$$
\cdot \left\{ \frac{1}{1 - (\mu - \alpha_1)\phi i} \dots + \frac{1}{1 - (\mu - \alpha_n)\phi i} \right\} .
$$

If a_k is replaced by its value $2(1 - \cos k\frac{2\pi}{n})$, a simple reflection now reveals that, for any even n, the function $\chi(\mu)$ assumes the same value for two values of μ, one of which is as smaller than 2 by the same amount as the other is larger. For if we put $\mu = 2 + \lambda$, we see at once that the interchange of λ and $-\lambda$ merely entails an interchange of the terms of the integrand. Thus for arbitrary even n the function $\chi(\mu)$ is symmetric about the value $\mu = 2$. It is easy to conclude that the same holds for odd n at least approximately, the more so the larger n is. A more detailed consideration of the integral regarding its numerical value for different values of the argument reveals moreover that the function $\chi(\mu)$ attains an absolute maximum for $\mu = 2$ (strictly for all even n and to a close approximation for odd n). From this maximum the function decreases rapidly in both directions as soon as n is reasonably large. We state these results without proof.

These properties lead to the conclusion that the probability of occurrence of a set of errors for which the ratio μ lies below a definite limit $(2 - \lambda)$ is equal to the probability that μ exceeds the value $(2 + \lambda)$. We denote the former by $X(2 - \lambda)$ and the latter by $1 - X(2 + \lambda)$. Moreover, their common value decreases to zero for increasing λ, the more rapidly the larger n. The probability that μ differs from 2 in absolute value by more than λ, evidently given by the expression

$$1 - (X(2 + \lambda) - X(2 - \lambda)),$$

consequently declines rapidly for increasing n if λ is appreciably different from zero.

Now the value attained by the ratio $\mu = \Theta/\Delta$ in a particular case is, in view of the original definition of the function Θ, largely determined by the pattern of the distribution of the errors. It follows, to start with in general outline, that the above implies a certain conformity to a law controlling this distribution as long as the errors originate from unsystematic causes. This conformity may evidently be described as follows: for errors of random origin those sets appear most frequently for which the ratio of the sum of the squared differences to the sum of squares of the errors $= 2$. Those sets for which this ratio is appreciably different from 2 and approaches the limits 0 and 4 appear relatively very seldom. Or, since

$$\Theta = 2\Delta - 2(x_1 x_2 + x_2 x_3 \ldots + x_{n-1} x_n + x_n x_1),$$

i.e.,

$$\lambda = \pm(\mu - 2) = 2(z_1 x_2 + x_2 x_3 \ldots + x_n x_1)/\Delta,$$

it follows that those patterns of error appear most frequently for which the algebraic sum of the products of two successive errors is zero.

The point just made gives rise to a simple criterion for judging whether any given set of errors originates from sources of error that act irregularly. Such a judgment needs to be based on the investigation of the value, assumed for the given set of errors, by the ratio $\Theta/\Delta = \mu$ or by the equivalent $2\Theta/\Delta = \lambda$ (where the sum of products of neighboring terms is denoted by θ). The absolute value of the difference $(\mu - 2)$ or the absolute value of λ will be the determining factor on which the probability P of a random origin depends. If that value differs substantially from zero, then P will be the smaller, the larger the set of errors, i.e., the more extensive the underlying series of observations. In such a case we must conclude with all the more probability that systematic sources of error are influencing the observational process or that there is an actual departure from the underlying distributional assumptions.

The numerical determination of the probability level of the one or the other explanation [i.e., random or nonrandom source of error (translator)] depends, of course, on the numerical value of the function $X(\mu)$ for arbitrary values of the argument. We require, therefore, a representation of

$X(\mu)$ in a form convenient for numerical evaluation. This can be done without difficulty in several ways, at least under the assumption of n large, the only case of interest in practice. One way follows immediately from the definite integral in equation (20), another through a suitable transformation of the summation expression to which the integral can be reduced. However, the execution must be omitted here, as the preparations still needed for this purpose would too greatly exceed the space of this treatise.

The Distribution of the Sample Variance Under Normality

Comments on Helmert (1876b)

1 Introduction

The article translated here is actually just one part, with its own title, of a three-part paper that slays several dragons (Helmert, 1876b). Let $X_1, \ldots, X_n$ be independent $N(\mu, \sigma^2)$ variates. In the omitted portions Helmert derives essentially the variance of the mean deviation $\frac{1}{n} \sum |X_i - \bar{X}|$ and of the mean difference $\frac{1}{n(n-1)} \sum_{i \neq j} |X_i - X_j|$.

Helmert, an eminent geodesist, is the first to show, in effect, that

$$\sum (X_i - \bar{X})^2 / \sigma^2 \sim \chi^2_{n-1}. \tag{a}$$

To do this he introduces two transformations that in combination are known as the Helmert transformation. As we will see, there are also other points of interest in the selected extract.

Helmert's notation has been modified somewhat, being particularly confusing to the modern reader (beyond his use of μ for the standard deviation, shared by Gauss and others). All carets denoting estimates have been added for clarity. In the title of the article, the "adjusted direct observations" are the deviations $X_i - \bar{X}$, $i = 1, \ldots, n$, denoted by $\lambda_i = \epsilon_i - \bar{\epsilon}$, where ϵ_i is the "true error" $X_i - \mu$. Following Gauss, Helmert writes $[\lambda\lambda]$ for $\sum \lambda_i^2$. He works with both the "precision" $h = 1/\sigma\sqrt{2}$ and σ, the "mean error." The "probable error" ρ, long in use, is given by

$$P(-\rho\sigma < \epsilon < \rho\sigma) = \frac{1}{2};$$

i.e., $\rho = \sigma\Phi^{-1}(0.75) = 0.67449\ldots\sigma$, where Φ denotes the standard normal cdf.

A section of the paper dealing with Fechner's formula, referred to in the title, has been eliminated, as have some other minor passages.

2   The χ^2 Distribution

We turn to the derivation of (a). The language surrounding (2), relating to prior and posterior distributions, is no longer familiar. Influenced by Gauss (1816) Helmert uses, in obvious notation,

$$f(h|\epsilon) = \frac{f(\epsilon|h)f(h)}{f(\epsilon)} \propto f(\epsilon|h),$$

h being assumed uniformly distributed. He then maximizes the "future probability" $f(\epsilon|h)d\epsilon$ w.r.t. h to obtain the posterior mode (which with h uniform is also the MLE). This involves the unknown quantity $[\epsilon\epsilon]$ in [A] (our labeling) and motivates Helmert to determine the distribution of the observable $[\lambda\lambda]$.

The transformation in Section 1 shows the modern reader immediately that sample mean and variance are stochastically independent, but Helmert fails to notice this. He simply integrates over $\bar{\epsilon}$ to obtain what he needs, the joint distribution of the λ's. Helmert now applies a second very ingenious transformation (4). A likely explanation of how he arrived at this transformation is given by Hald (1952, p. 267ff.) It is the combination of these two transformations, together with $t_n = \sqrt{n}\,\bar{\epsilon}$ and the relabeling $\epsilon_n \to \epsilon_1$, $\epsilon_j \to \epsilon_{j+1}$, $j = 1, \ldots, n-1$, that results in the useful orthogonal transformation involving $n-1$ successive linear contrasts

$$t_j = [j(j+1)]^{1/2}(j\epsilon_{j+1} - \epsilon_1 - \cdots - \epsilon_j)$$

and

$$t_n = (\epsilon_1 + \cdots + \epsilon_n)/\sqrt{n}$$

that has become known as a Helmert transformation. As Hald (2000) points out, it was Pizzetti (1891, p. 129) who first displayed the Helmert transformation.

Since $\sum_{i=1}^{n}(\epsilon_i - \bar{\epsilon})^2 = \sum_{j=1}^{n-1} t_j^2$, the desired result follows immediately from Helmert's proof of

$$\sum(X_i - \mu)^2/\sigma^2 \sim \chi_n^2 \tag{b}$$

announced in 1875 and established by induction in Helmert (1876a). He was evidently unaware of earlier proofs of (b) by Bienaymé (1852) and Abbe (1863); see David (1998).

Moreover, Helmert did not realize that in eq. (B) of his paper he had proved that $\bar{\epsilon} \sim N(0, (2nh^2)^{-1})$, a result he laboriously obtained from scratch in an earlier untranslated portion of the paper. In any case, the normality of $\bar{\epsilon}$ can not have been unknown (see Hald, 1998, p. 634 ff.).

Helmert's lack of familiarity with previous relevant literature is repaid in kind by Karl Pearson and R.A. Fisher. Although Helmert's work receives detailed (uncritical) attention in a major text by Czuber (1891) and

Pearson knew German well, there is no reference to Helmert in Pearson (1900), the famous paper in which he "discovers" and names the χ^2 distribution (but in a goodness-of-fit context). In Pearson (1931) he apologizes for Helmert's work having been overlooked by the "English school of statisticians" and suggests that the distribution should in future be named for Helmert. It seems that Fisher became aware of Helmert only then, having swiftly derived the distribution of $\sum(X_i - \bar{X})^2/\sigma^2$ by a geometric argument himself *and* established the independence of $\sum(X_i - \bar{X})^2$ and $\bar{X}$ in Fisher (1920).

3   The Estimation of σ^2 and σ

As part of more general results in the theory of least squares, Gauss (1823) shows that for any population (with finite variance)

$$\hat{\sigma}^2 = \sum(X_i - \bar{X})^2/(n-1)$$

is an unbiased estimator of σ^2. He even obtains expressions for the variance of $\hat{\sigma}^2$ which in the normal case reduce to

$$\mathrm{var}(\hat{\sigma}^2) = 2\sigma^4/(n-1) \qquad n > 1. \tag{c}$$

In Section 2 Helmert apparently thinks that he has established $\hat{\sigma}^2$ as the posterior mode (or MLE) of σ^2. What he has actually shown is exactly what he states: for given values of the differences $X_i - \bar{X}$, the MLE of σ^2 is $\hat{\sigma}^2$. But usually we are given the values of the X_i, in which case the MLE is, of course, the biased estimator $\sum(X_i - \bar{X})^2/n$. We might add that given only the value of $\sum(X_i - \bar{X})^2$, the MLE is $\sum(X_i - \bar{X})^2/(n-3)$. There is no evidence that Helmert actually followed up on his footnote (Sheynin, 1995).

To estimate σ Gauss simply uses $\hat{\sigma}$, as does Helmert in eq. (1). The estimated standard error formula in (1) is just the large-sample result corresponding to (c), as Helmert states in Section 4. As a formula for n large, (1) is unexceptional, but Helmert wants to find a standard error formula valid for small n also. In fact, it is this aim that more specifically leads him to find the distribution of $\sum(X_i - \bar{X})^2$. However, although using this distribution to evaluate $E[\sum(X_i - \bar{X})^2]^{1/2}$, Helmert fails to realize in (6) that $E(\hat{\sigma}) \neq \sigma$ for $n < \infty$. The basic problem seems to be that estimation procedures developed for symmetrical distributions are used without change for the asymmetrical distribution of $\hat{\sigma}$.

4   Concluding Comments

This paper must leave one smiling, with a sense of wonder: the distribution of $\sum(X_i - \bar{X})^2 = (n-1)S^2$ under normality is brilliantly obtained as a byproduct of a flawed investigation! Helmert sets out to find the exact variance of S, but is convinced that S is unbiased for σ, in spite of generating evidence to the contrary. He fails to notice that he has established the independence of $\bar{X}$ and S^2 and the normality of $\bar{X}$ (the latter not a new result). But Helmert admirably overcomes a lack of background knowledge by rederiving elsewhere the normal distribution of $\bar{X}$ and the χ_n^2 distribution of $\sum(X_i - \mu)^2$. It is entirely fitting that his shortcomings are forgotten and his brilliance remembered.

For a good account of many aspects of this paper, see also Hald (1998, pp. 635–637). An extensive review of Helmert's work in statistics is given by Sheynin (1995). Both authors provide some biographical details.

References

Abbe, E. (1863). Ueber die Gesetzmässigkeit in der Vertheilung der Fehler bei Beobachtungsreihen. *Gesammelte Abhandlungen*, Vol. 2. Gustav Fischer, Jena.

Bienaymé, I.-J. (1852). Sur la probabilité des erreurs d'après la méthode des moindres carrés. *J. math. pures et appliquées*, **17**, 33–78.

Czuber, E. (1891). *Theorie der Beobachtungsfehler*. Teubner, Leipzig.

David, H.A. (1998). Early measures of variability. *Statist. Sci.*, **13**, 368–377.

Fisher, R.A. (1920). A mathematical examination of the methods of determining the accuracy of an observation by the mean error, and by the mean square error. *Mon. Not. R. Astron. Soc.*, **80**, 758–770. [Reprinted as Paper 2 in Fisher (1950).]

Fisher, R.A. (1950). *Contributions to Mathematical Statistics*. Wiley, New York.

Gauss, C.F. (1816). Bestimmung der Genauigkeit der Beobachtungen. In *Carl Friedrich Gauss Werke*, Vol. 4. Göttingen: Königliche Gesellschaft der Wissenschaften, 1880, pp. 109–117.

Gauss, C.F. (1821, 1823, 1826). Theoria combinationis observationum erroribus minimis obnoxiae. In *Carl Friedrich Gauss Werke*, Vol. 4. Göttingen: Königliche Gesellschaft der Wissenschaften, 1880, pp. 3–93. [English translation by Stewart (1995).]

Hald, A. (1952). *Statistical Theory with Engineering Applications*. Wiley, New York.

Hald, A. (1998). *A History of Mathematical Statistics from 1750 to 1930*. Wiley, New York.

Hald, A. (2000). Pizzetti's contributions to the statistical analysis of normally distributed observations, 1891. *Biometrika*, **87**, 213–217.

Helmert, F.R. (1875). Ueber die Berechnung des wahrscheinlichen Fehlers aus einer endlichen Anzahl wahrer Beobachtungsfehler. *Zeit. Math. u. Phys.*, **20**, 300–303.

Helmert, F.R. (1876a). Ueber die Wahrscheinlichkeit der Potenzsummen der Beobachtungsfehler und über einige damit im Zusammenhange stehende Fragen. *Zeit. Math. u. Phys.*, **21**, 192–218.

Helmert, F.R. (1876b). Die Genauigkeit der Formel von Peters zur Berechnung des wahrscheinlichen Fehlers directer Beobachtungen gleicher Genauigkeit. *Astron. Nachr.*, **88**, 113–132.

Pearson, K. (1900). On the criterion that a given system of deviations from the probable in the case of a correlated system of variables is such that it can be reasonably supposed to have arisen from random sampling. *Phil. Mag.*, 5th Series, **50**, 157–175.

Pearson, K. (1931). Historical note on the distribution of the standard deviations of samples of any size drawn from an indefinitely large normal parent population. *Biometrika*, **23**, 416–418.

Pizzetti, P. (1891). I fundamenti matematici per la critica dei resultati sperimentali. Genoa. Reprinted as Vol. 3 in *Biblioteca di "Statistica"* (1963). "Statistica," Bologna. Page reference is to the 1963 reprint.

Sheynin, O. (1995). Helmert's work in the theory of errors. *Arch. Hist. Exact Sci.*, **49**, 73–104.

Stewart, G.W. (1995). *Theory of the Combination of Observations Least Subject to Errors*. SIAM, Philadelphia. [Translation of Gauss (1821, 1823, 1826).]

The Calculation of the Probable Error from the Squares of the Adjusted Direct Observations of Equal Precision and Fechner's Formula

F. R. Helmert

Let λ denote the deviations of the observations from their arithmetic mean, let σ denote the mean error, and ρ the probable error. Then the optimal estimate of ρ is well known to be given by the following formulae,

$$\hat{\rho} = 0.67449\ldots\hat{\sigma}$$

$$\hat{\sigma} = \sqrt{\frac{[\lambda\lambda]}{n-1}}\left[1 \pm \sqrt{\frac{1}{2(n-1)}}\right],$$

(1)

where the square root in the bracket is the mean error in the estimate of $\hat{\sigma}$, expressed as a fraction of $\hat{\sigma}$. It is our intention to provide a somewhat more rigorous derivation of this formula under the Gaussian law of error than given elsewhere, even where the principles of probability theory are used.

If ϵ denotes a true error of an observation, then the future probability of a set $\epsilon_1, \ldots, \epsilon_n$ is

$$\left[\frac{h}{\sqrt{\pi}}\right]^n e^{-h^2[\epsilon\epsilon]} d\epsilon_1 \ldots d\epsilon_n .$$

(2)

For given $\epsilon_1, \ldots, \epsilon_n$, by setting the probability of a hypothesis about h proportional to this expression, one obtains as optimal value of σ^2

$$\frac{1}{2\hat{h}^2} = \hat{\sigma}^2 = \frac{[\epsilon\epsilon]}{n} .$$

(A)

However, since the ϵ are unknown, we are forced to estimate $[\epsilon\epsilon]$, and this may be regarded as a weakness of previous derivations. This deficiency may be removed by the consideration that a set $\lambda_1, \ldots, \lambda_n$ may arise from true errors in an infinity of ways. But since only the λ are given, we must calculate the future probability of a set $\lambda_1, \ldots, \lambda_n$ and take this expression as proportional to the probability of the hypothesis about h.

1   Probability of a Set $\lambda_1, \ldots, \lambda_n$ of Deviations from the Arithmetic Mean

In expression (2) we introduce the variables $\lambda_1, \ldots, \lambda_{n-1}$ and $\bar{\epsilon}$ in place of the ϵ by the equations:

$$\epsilon_1 = \lambda_1 + \bar{\epsilon}, \quad \epsilon_2 = \lambda_2 + \bar{\epsilon}, \ldots$$

$$\epsilon_{n-1} = \lambda_{n-1} + \bar{\epsilon}, \quad \epsilon_n = -\lambda_1 - \lambda_2 - \cdots - \lambda_{n-1} + \bar{\epsilon}.$$

This transformation is in accord with the known relations between true errors ϵ and deviations λ, since the addition of the equations gives $n\bar{\epsilon} = [\epsilon]$; at the same time the condition $[\lambda] = 0$ is satisfied. The determinant of the transformation, a determinant of the nth degree, is

$$\begin{vmatrix} 1 & \cdot & \cdot & & \cdot & 1 \\ \cdot & 1 & \cdot & & \cdot & 1 \\ \cdot & \cdot & 1 & & \cdot & 1 \\ & & & & & \\ \cdot & \cdot & \cdot & & 1 & 1 \\ -1 & -1 & -1 & & -1 & 1 \end{vmatrix} = n \ .$$

Consequently expression (2) becomes

$$n \left[\frac{h}{\sqrt{\pi}} \right]^n e^{-h^2[\lambda\lambda] + h^2 n\bar{\epsilon}^2} d\lambda_1 d\lambda_2 \ldots d\lambda_{n-1} d\bar{\epsilon} \ , \tag{B}$$

where $[\lambda\lambda] = \lambda_1^2 + \lambda_2^2 + \cdots + \lambda_n^2$; $\lambda_n = -\lambda_1 - \lambda_2 - \cdots - \lambda_{n-1}$. If we now integrate over all possible values of $\bar{\epsilon}$, we obtain for the probability of the set $\lambda_1 \cdots \lambda_n$ the expression

$$\sqrt{n} \left[\frac{h}{\sqrt{\pi}} \right]^{n-1} e^{-h^2[\lambda\lambda]} d\lambda_1 \ldots d\lambda_{n-1} \ . \tag{3}$$

This may be verified by integration over all possible values of $\lambda_1 \ldots \lambda_{n-1}$, which yields unity, as required.

2   Optimal Hypothesis on h for Given Deviations λ

For given values of the λ's we set the probability of a hypothesis on h proportional to expression (3). A standard argument then yields the optimal estimate of h as the value maximizing (3). Differentiation shows that this occurs when

$$\frac{1}{2h^2} = \frac{[\lambda\lambda]}{n-1} ,$$

which establishes the first part of formula (1)*.

3   Probability of a Sum $[\lambda\lambda]$ of Squares of the Deviations λ

The probability that $[\lambda\lambda]$ lies between u and $u + du$ is from (3)

$$\sqrt{n}\left[\frac{h}{\sqrt{\pi}}\right]^{n-1} \int d\lambda_1 \ldots \int d\lambda_{n-1}\, e^{-h^2[\lambda\lambda]} ,$$

integrated over all $\lambda_1 \ldots \lambda_{n-1}$ satisfying

$$u \leq [\lambda\lambda] \leq u + du .$$

We now introduce $n - 1$ new variables t by means of the equations:

$$
\begin{aligned}
t_1 &= \sqrt{2}\left(\lambda_1 + \tfrac{1}{2}\lambda_2 + \tfrac{1}{2}\lambda_3 + \tfrac{1}{2}\lambda_4 + \cdots + \tfrac{1}{2}\lambda_{n-1}\right) \\
t_2 &= \sqrt{\tfrac{3}{2}}\left(\lambda_2 + \tfrac{1}{3}\lambda_3 + \tfrac{1}{3}\lambda_4 + \cdots + \tfrac{1}{3}\lambda_{n-1}\right) \\
t_3 &= \sqrt{\tfrac{4}{3}}\left(\lambda_3 + \tfrac{1}{4}\lambda_4 + \cdots + \tfrac{1}{4}\lambda_{n-1}\right) \qquad (4) \\
\cdot\ & \qquad\qquad\qquad \cdot \qquad\quad \cdot \qquad\qquad\qquad\quad \cdot \\
t_{n-1} &= \sqrt{\tfrac{n}{n-1}}\lambda_{n-1} .
\end{aligned}
$$

With the determinant $\sqrt{n}$ of the transformation, the above expression becomes

$$\left[\frac{h}{\sqrt{\pi}}\right]^{n-1} \int dt_1 \ldots \int dt_{n-1}\, e^{-h^2[tt]} ,$$

the limits of integration being determined by the condition

$$u \leq [tt] \leq u + du .$$

---

*In the same way it is possible by strict use of probability theory to derive a formula for σ^2 when n observations depend on m unknowns, a result the author has established to his satisfaction and will communicate elsewhere.

We now recognize that the probability for the sum of squares of the n deviations λ, $[\lambda\lambda] = u$, is precisely the same as the probability that the sum of squares $[tt]$ of $n-1$ true errors equals u. This last probability I gave in Schlömilch's journal, 1875, p. 303, according to which

$$\frac{h^{n-1}}{\Gamma\left(\frac{n-1}{2}\right)}\, u^{\frac{n-3}{2}}\, e^{-h^2 u} du \ , \tag{5}$$

is the probability that the sum of squares $[\lambda\lambda]$ of the deviations λ of n equally precise observations from their mean lies between u and $u + du$. Integration of (5) from $u = 0$ to ∞ gives unity.

4 The Mean Error of the Formula $\hat{\sigma} = \sqrt{[\lambda\lambda] : (n-1)}$

Since it is difficult to obtain a generally valid formula for the probable error of this formula, we confine ourselves to the mean error.

The mean error of the formula $\hat{\sigma}^2 = \frac{[\lambda\lambda]}{n-1}$ is known exactly, namely $\sigma^2 \sqrt{2 : (n-1)}$. We have therefore

$$\hat{\sigma}^2 = \frac{[\lambda\lambda]}{n-1}\left[1 \pm \sqrt{\frac{2}{n-1}}\right]$$

and if n is large it follows by a familiar argument that

$$\hat{\sigma} = \sqrt{\frac{[\lambda\lambda]}{n-1}}\left[1 \pm \frac{1}{2}\sqrt{\frac{2}{n-1}}\right].$$

Formula (1) results. However, if n is small, for example equal to 2, this argument lacks all validity. For then $\sqrt{2 : (n-1)}$ is no longer small compared to 1, in fact even larger than 1 for $n = 2$. We now proceed as follows.

The mean squared error of the formula

$$\hat{\sigma} = \sqrt{[\lambda\lambda] : (n-1)}$$

is the mean value of

$$\left[\sqrt{\frac{[\lambda\lambda]}{n-1}} - \sigma\right]^2 .$$

If one develops the square and recalls that $[\lambda\lambda] : (n-1)$ has mean σ^2 or

$1 : 2h^2$, it follows that the mean of the above is

$$\frac{1}{h^2} - \frac{\sqrt{2}}{h} \left[\sqrt{\frac{[\lambda\lambda]}{n-1}} \right],$$

where the term in large brackets must be replaced by its mean value.

Consideration of formula (5) yields for the mean value of $\sqrt{[\lambda\lambda]}$ the expression

$$\frac{h^{n-1}}{\Gamma\left(\frac{n-1}{2}\right)} \int_0^\infty u^{\frac{n-2}{2}} e^{-h^2 u^2} du, \text{ i.e., } \frac{\Gamma\left(\frac{n}{2}\right)}{h\Gamma\left(\frac{n-1}{2}\right)},$$

so that the mean squared error of $\hat{\sigma}$ is

$$\frac{1}{h^2} \left[1 - \frac{\Gamma\left(\frac{n}{2}\right)}{\Gamma\left(\frac{n-1}{2}\right)} \sqrt{\frac{2}{n-1}} \right].$$

We must therefore regard the following formula as more accurate than (1):

$$\hat{\sigma} = \sqrt{\frac{[\lambda\lambda]}{n-1}} \left[1 \pm \sqrt{\left\{ 2 - \frac{\Gamma\left(\frac{n}{2}\right)}{\Gamma\left(\frac{n-1}{2}\right)} \sqrt{\frac{8}{n-1}} \right\}} \right]$$

$$\hat{\rho} = 0.67449\ldots\hat{\sigma}, \tag{6}$$

where the square root following $\pm$ signifies the mean error of the formula for $\hat{\sigma}$.

The Random Walk and Its Fractal Limiting Form

Comments on Venn (1888)

1 Introduction

John Venn, whose name is immortalized in the "Venn diagram" of logic and set theory (1880) which replaced the more confusing "Euler diagram" previously used for the same purpose, published *The Logic of Chance* in 1866. The third edition, of 1888, contained much new material, including a chapter "*The conception of randomness*," in which Venn illustrated the "truly random character" of the digits of π by using them to generate a discrete random walk in two dimensions. He then discussed the limiting form that such a random walk would take if the direction were drawn from a uniform distribution and the step length made indefinitely small, describing the result in terms which we would now encapsulate in the word "fractal."

Venn constructs the discrete random walk on a square lattice by allowing jumps to be made from a point to each of its eight neighbors reachable with one step horizontally, vertically, or diagonally. The eight directions are assigned equal probabilities generated by the digits 0 to 7 in π. Taking one of the directions (say the horizontal) he observes that the resulting distribution is binomial (without actually using the word, in keeping with the tenor of his book in which he is always at pains to suppress any technical material) and he describes the bivariate case. We can be in little doubt that for the limiting case Venn was aware that he was dealing with a bivariate normal distribution, for he was of course perfectly familiar with the normal as the limiting form of the binomial (*The Logic of Chance*, p. 36).

Although the notion of a random walk is implicit in the earliest modern writings on probability from the time of Pascal (see, for example, Edwards, 1987), it is only during the last hundred years that it has come to be the preferred description for many stochastic processes, and Venn's figure, with which his chapter ends, may be the first published diagram of a random walk.

The continuous limiting-case random walk was soon to become famous as the simple diffusion model of physics, the connection with physics being initially made through some amusing correspondence in the scientific journal *Nature*. Karl Pearson (1905a) wrote a letter under the heading "The problem of the random walk" (the first occurrence of the phrase) in which he described a walk with random direction and discrete steps of equal length and enquired whether any reader could refer him to a work which would tell him the probability distribution of the total distance travelled. Lord Rayleigh was quick to point out (in the following week's issue; Rayleigh, 1905) that he had treated the problem in connection with vibrations of unit amplitude and arbitrary phase in papers in 1880 and 1899. The next week's issue contained Pearson's admission that he "ought to have known" Rayleigh's solution (Pearson, 1905b). Rayleigh (1880) did indeed use the analogy of a random walk, though without either that phrase or a diagram: "It will be convenient in what follows to consider the vibrations to be represented by lines (of unit length) drawn from a fixed point O, the intersection of rectangular axes Ox and Oy."

2   A Note on the Author

Venn was born in Hull, England, in 1834 and died in Cambridge in 1923. He was Lecturer in Moral Science (that is, Philosophy) at Gonville and Caius College, Cambridge, from 1862, a Fellow of the College, and ultimately President. As well as *The Logic of Chance* he wrote *Symbolic Logic* (1881) in which his logic diagrams featured prominently, and *Empirical Logic* (1889). *The Logic of Chance* was influential in turning statistical opinion away from a Bayesian formulation towards "repeated-sampling" methods, and in particular it influenced R.A. Fisher, who was a young Fellow of Gonville and Caius College when Venn was President.

Venn gave the first lecture course in the "Theory of Statistics" in England as part of the Cambridge University Moral Science Tripos in 1890. In later life he developed an interest in Cambridge college and university history, and the *Biographical History* of his college (Volume I, 1897) is acclaimed by historians as the first example of its kind, in which brief lives of all the members are recorded.

References

Edwards, A.W.F. (1987). *Pascal's Arithmetical Triangle*. Griffin, London and Oxford University Press, New York.

Pearson, K. (1905a). The problem of the random walk. *Nature*, **72**, 294.

Pearson, K. (1905b). The problem of the random walk. *Nature*, **72**, 342.

Rayleigh, Lord. (1880). On the resultant of a large number of vibrations of the same pitch and arbitrary phase. *Phil. Mag.*, **10**, 73–78.

Rayleigh, Lord. (1905). The problem of the random walk. *Nature*, **72**, 318.

Venn, J. (1888). *The Logic of Chance*, 3rd edn. Macmillan, London.

§ 10.   III.   Apart from definitions and what comes of them, perhaps the most important question connected with the conception of Randomness is this: How in any given case are we to determine whether an observed arrangement is to be considered a random one or not? This question will have to be more fully discussed in a future chapter, but we are already in a position to see our way through some of the difficulties involved in it.

(1)   If the events or objects under consideration are supposed to be continued indefinitely, or if we know enough about the mode in which they are brought about to detect their ultimate tendency,—or even, short of this, if they are numerous enough to be beyond practical counting,—there is no great difficulty. We are simply confronted with a question of fact, to be settled like other questions of fact. In the case of the rain-drops, watch two equal squares of pavement or other surfaces, and note whether they come to be more and more densely uniformly and evenly spotted over: if they do, then the arrangement is what we call a random one. If I want to know whether a tobacco-pipe really breaks at random, and would therefore serve as an illustration of the problem proposed some pages back, I have only to drop enough of them and see whether pieces of all possible lengths are equally represented in the long run. Or, I may argue deductively, from what I know about the strength of materials and the molecular constitution of such bodies, as to whether fractures of small and large pieces are all equally likely to occur.

§ 11.   The reader's attention must be carefully directed to a source of confusion here, arising out of a certain cross-division. What we are now discussing is a question of fact, viz. the nature of a certain ultimate arrangement; we are not discussing the particular way in which it is brought about. In other words, the antithesis is between what is and what is not random: it is not between what is random and what is designed. As we shall see in a few moments it is quite possible that an arrangement which is the result,—if ever anything were so,—of 'design', may nevertheless present the unmistakeable stamp of randomness of arrangement.

Consider a case which has been a good deal discussed, and to which we shall revert again: the arrangement of the stars. The question here is rather complicated by the fact that we know nothing about the actual mutual positions of the stars, all that we can take cognizance of being their apparent or visible places as projected upon the surface of a supposed sphere. Appealing to what alone we can thus observe, it is obvious that the arrangement, as a whole, is not of the random sort. The Milky Way and the other resolvable nebulæ, as they present themselves to us, are as obvious an infraction of such an arrangement as would be the occurrence here and there of patches of ground in a rain-fall which received a vast number more drops than the spaces surrounding them. If we leave these exceptional areas out of the question and consider only the stars which are visible by the naked eye or by slight telescopic power, it seems equally certain that the arrangement *is*, for the most part, a fairly representative random one. By this we mean nothing more than the fact that when we mark off any number of equal areas on the visible sphere these are found to contain approximately the same number of stars.

The actual arrangement of the stars in space *may* also be of the same character: that is, the apparently denser aggregation may be apparent only, arising from the fact that we are looking through regions which are not more thickly occupied but are merely more extensive. The alternative before us, in fact, is this. If the whole volume, so to say, of the starry heavens is tolerably regular in shape, then the arrangement of the stars is not of the random order; if that volume is very irregular in shape, it is possible that the arrangement within it may be throughout of that order.

§ 12. (2) When the arrangement in question includes but a comparatively small number of events or objects, it becomes much more difficult to determine whether or not it is to be designated a random one. In fact we have to shift our ground, and to decide not by what has been actually observed but by what we have reason to conclude would be observed if we could continue our observation much longer. This introduces what is called 'Inverse Probability', viz. the determination of the nature of a cause from the nature of the observed effect; a question which will be fully discussed in a future chapter. But some introductory remarks may be conveniently made here.

Every problem of Probability, as the subject is here understood, introduces the conception of an ultimate limit, and therefore presupposes an indefinite possibility of repetition. When we have only a finite number of occurrences before us, *direct* evidence of the character of their arrangement fails us, and we have to fall back upon the nature of the agency which produces them. And as the number becomes smaller the confidence with which we can estimate the nature of the agency becomes gradually less.

Begin with an intermediate case. There is a small lawn, sprinkled over with daisies: is this a random arrangement? We feel some confidence that it is so, on mere inspection; meaning by this that (negatively) no trace of any regular pattern can be discerned and (affirmatively) that if we take any moderately small area, say a square yard, we shall find much about the same number of the plants included in it. But we can help ourselves by an appeal to the known agency of distribution here. We know that the daisy spreads by seed, and considering the effect of the wind and the continued sweeping and mowing of the lawn we can detect causes at work which are analogous to those by which the dealing of cards and the tossing of dice are regulated.

In the above case the appeal to the process of production was subsidiary, but when we come to consider the nature of a very small succession or group this appeal becomes much more important. Let us be told of a certain succession of 'heads' and 'tails' to the number of ten. The range here is far too small for decision, and unless we are told whether the agent who obtained them was tossing or designing we are quite unable to say whether or not the designation of 'random' ought to be applied to the result obtained. The truth must never be forgotten that though 'design' is sure to break down in the long run if it make the attempt to produce directly the semblance of randomness[1], yet for a short spell it can simulate it perfectly. Any short succession, say of heads and tails, may have been equally well brought about by tossing or by deliberate choice.

§ 13. The reader will observe that this question of randomness is being here treated as simply one of ultimate statistical fact. I have fully admitted that this is not the primitive conception, nor is it the popular interpretation, but to adopt it seems the only course open to us if we are to draw inferences such as those contemplated in Probability. When we look to the producing agency of the ultimate arrangement we may find this very various. It may prove itself to be (a few stages back) one of conscious deliberate

[1] Vide p. 68.

purpose, as in drawing a card or tossing a die: it may be the outcome of an extremely complicated interaction of many natural causes, as in the arrangement of the flowers scattered over a lawn or meadow: it may be of a kind of which we know literally nothing whatever, as in the case of the actual arrangement of the stars relatively to each other.

This was the state of things had in view when it was said a few pages back that randomness and design would result in something of a cross-division. Plenty of arrangements in which design had a hand, a stage or two back, can be mentioned, which would be quite indistinguishable in their results from those in which no design whatever could be traced. Perhaps the most striking case in point here is to be found in the arrangement of the digits in one of the natural arithmetical constants, such as π or ϵ, or in a table of logarithms. If we look to the process of production of these digits, no extremer instance can be found of what we mean by the antithesis of randomness: every figure has its necessarily pre-ordained position, and a moment's flagging of intention would defeat the whole purpose of the calculator. And yet, if we look to results only, no better instance can be found than one of these rows of digits if it were intended to illustrate what we practically understand by a chance arrangement of a number of objects. Each digit occurs approximately equally often, and this tendency developes as we advance further; the mutual juxtaposition of the digits also shows the same tendency, that is, any digit (say 5) is just as often followed by 6 or 7 as by any of the others. In fact, if we were to take the whole row of hitherto calculated figures, cut off the first five as familiar to us all, and contemplate the rest, no one would have the slightest reason to suppose that these had not come out as the results of a die with ten equal faces.

§ 14. If it be asked *why* this is so, a rather puzzling question is raised. Wherever physical causation is involved we are generally understood to have satisfied the demand implied in this question if we assign antecedents which will be followed regularly by the event before us; but in geometry and arithmetic there is no opening for antecedents. What we then commonly look for is a demonstration, i.e. the re-solution of the observed fact into axioms if possible, or at any rate into admitted truths of wider generality. I do not know that a demonstration can be given as to the existence of this characteristic of statistical randomness in such successions of digits as those under consideration. But the following remarks may serve to shift the onus of unlikeli-hood by suggesting that the preponderance of analogy is rather in favour of the existence.

Take the well-known constant π for consideration. This stands for a quantity which presents itself in a vast number of arithmetical and geometrical relations; let us take for examination the best known of these, by regarding it as standing for the ratio of the circumference to the diameter of a circle. So regarded, it is nothing more than a simple case of the measurement of a magnitude by an arbitrarily selected unit. Conceive then that we had before us a rod or line and that we wished to measure it with absolute accuracy. We must suppose—if we are to have a suitable analogue to the determination of π to several hundred figures,—that by the application of continued higher magni-fying power we can detect ever finer subdivisions in the graduation. We lay our rod against the scale and find it, say, fall between 31 and 32 inches; we then look at the next division of the scale, viz. that into tenths of an inch. Can we see the slightest reason why the number of these tenths should be other than independent of the number of

123

whole inches? The "piece over" which we are measuring may in fact be regarded as an entirely new piece, which had fallen into our hands after that of 31 inches had been measured and done with; and similarly with every successive piece over, as we proceed to the ever finer and finer divisions.

Similar remarks may be made about most other incommensurable quantities, such as irreducible roots. Conceive two straight lines at right angles, and that we lay off a certain number of inches along each of these from the point of intersection; say two and five inches, and join the extremities of these so as to form the diagonal of a right-angled triangle. If we proceed to measure this diagonal in terms of either of the other lines we are to all intents and purposes extracting a square root. We should expect, rather than otherwise, to find here, as in the case of π, that incommensurability and resultant randomness of order in the digits was the rule, and commensurability was the exception. Now and then, as when the two sides were three and four, we should find the diagonal commensurable with them; but these would be the occasional exceptions, or rather they would be the comparatively finite exceptions amidst the indefinitely numerous cases which furnished the rule.

§ 15. The best way perhaps of illustrating the truly random character of such a row of figures is by appealing to graphical aid. It is not easy here, any more than in ordinary statistics, to grasp the import of mere figures; whereas the arrangement of groups of points or lines is much more readily seized. The eye is very quick in detecting any symptoms of regularity in the arrangement, or any tendency to denser aggregation in one direction than in another. How then are we to dispose our figures so as to force them to display their true character? I should suggest that we set about *drawing a line at random;* and, since we cannot

V.   8 trust our own unaided efforts to do this, that we rely upon the help of such a table of figures to do it for us, and then examine with what sort of efficiency they can perform the task. The problem of drawing *straight* lines at random, under various limitations of direction or intersection, is familiar enough, but I do not know that any one has suggested the drawing of a line whose shape as well as position shall be of a purely random character. For simplicity we suppose the line to be confined to a plane.

The definition of such a line does not seem to involve any particular difficulty. Phrased in accordance with the ordinary language we should describe it as the path (i.e. any path) traced out by a point which at every moment is as likely to move in any one direction as in any other. That we could not ourselves draw such a line, and that we could not get it traced by any physical agency, is certain. The mere inertia of any moving body will always give it a tendency, however slight, to go on in a straight line at each moment, instead of being instantly responsive to instantaneously varying dictates as to its direction of motion. Nor can we conceive or picture such a line in its ultimate or ideal condition. But it is easy to give a graphical approximation to it, and it is easy also to show how this approximation may be carried on as far as we please towards the ideal in question.

We may proceed as follows. Take a sheet of the ordinary ruled paper prepared for the graphical exposition of curves. Select as our starting point the intersection of two of these lines, and consider the eight 'points of the compass' indicated by these lines and the bisections of the contained right angles [1]. For suggesting the random selection amongst

[1] It would of course be more complete to take *ten* alternatives of direction, and thus to omit none of the digits; but this is much more troublesome in practice than to confine ourselves to eight.

these directions let them be numbered from 0 to 7, and let us say that a line measured due 'north' shall be designated by the figure 0, 'north-east' by 1, and so on. The selection amongst these numbers, and therefore directions, at every corner, might be handed over to a die with eight faces; but for the purpose of the illustration in view we select the digits 0 to 7 as they present themselves in the calculated value of π. The sort of path along which we should travel by a series of such steps thus taken at random may be readily conceived; it is given at the end of this chapter.

For the purpose with which this illustration was proposed, viz. the graphical display of the succession of digits in any one of the incommensurable constants of arithmetic or geometry, the above may suffice. After actually testing some of them in this way they seem to me, so far as the eye, or the theoretical principles to be presently mentioned, are any guide, to answer quite fairly to the description of randomness.

§ 16. As we are on the subject, however, it seems worth going farther by enquiring how near we could get to the ideal of randomness of direction. To carry this out completely two improvements must be made. For one thing, instead of confining ourselves to eight directions we must admit an infinite number. This would offer no great difficulty; for instead of employing a small number of digits we should merely have to use some kind of circular teetotum which would rest indifferently in any direction. But in the next place instead of short finite steps we must suppose them indefinitely short. It is here that the actual unattainability makes itself felt. We are familiar enough with the device, employed by Newton, of passing from the discontinuous polygon to the continuous curve. But we can resort to this device because the ideal, viz. the curve, is as easily drawn (and, I should say, as easily conceived or pictured) as any of the steps which lead us towards it. But in the case before us it is otherwise. The line in question will remain discontinuous, or rather angular, to the last: for its angles do not tend even to lose their sharpness, though the fragments which compose them increase in number and diminish in magnitude without any limit. And such an ideal is not conceivable as an ideal. It is as if we had a rough body under the microscope, and found that as we subjected it to higher and higher powers there was no tendency for the angles to round themselves off. Our 'random line' must remain as 'spiky' as ever, though the size of its spikes of course diminishes without any limit.

The case therefore seems to be this. It is easy, in words, to indicate the conception by speaking of a line which at every instant is as likely to take one direction as another. It is easy moreover to draw such a line with any degree of minuteness which we choose to demand. But it is not possible to conceive or picture the line in its ultimate form [1]. There is in fact no 'limit' here, intelligible to the understanding or picturable by the imagination (corresponding to the asymptote of a curve, or the continuous curve to the incessantly developing polygon), towards which we find ourselves continually approaching, and which therefore we are apt to conceive ourselves as ultimately attaining The usual assumption therefore which underlies the Newtonian infinitesimal geometry and the Differential Calculus, ceases to apply here.

§ 17. If we like to consider such a line in one of its approximate stages, as above indicated, it seems to me that

[1] Any more than we picture the shape of an equiangular spiral *at the centre*.

some of the usual theorems of Probability, where large numbers are concerned, may safely be applied. If it be asked, for instance, whether such a line will ultimately tend to stray indefinitely far from its starting point, Bernoulli's 'Law of Large Numbers' may be appealed to, in virtue of which we should say that it was excessively unlikely that its divergence should be relatively great. Recur to our graphical illustration, and consider first the resultant deviation of the point (after a great many steps) right or left of the vertical line through the starting point. Of the eight admissible motions at each stage two will not affect this relative position, whilst the other six are equally likely to move us a step to the right or to the left. Our resultant 'drift' therefore to the right or left will be analogous to the resultant difference between the number of heads and tails after a great many tosses of a penny. Now the well-known outcome of such a number of tosses is that ultimately the *proportional* approximation to the à priori probability, i.e. to equality of heads and tails, is more and more nearly carried out, but that the *absolute* deflection is more and more widely displayed.

Applying this to the case in point, and remembering that the results apply equally to the horizontal and vertical directions, we should say that after any very great number of such 'steps' as those contemplated, the ratio of our distance from the starting point to the whole distance travelled will pretty certainly be small, whereas the actual distance from it would be large. We should also say that the longer we continued to produce such a line the more pronounced would these tendencies become. So far as concerns this test, and that afforded by the general appearance of the lines drawn,—this last, as above remarked, being tolerably trustworthy,—I feel no doubt as to the generally 'random'

118   *Randomness and its scientific treatment.*   [CHAP. V.

character of the rows of figures displayed by the incommensurable or irrational ratios in question.

As it may interest the reader to see an actual specimen of such a path I append one representing the arrangement of the eight digits from 0 to 7 in the value of π. The data are taken from Mr Shanks' astonishing performance in the calculation of this constant to 707 places of figures (*Proc. of R. S.*, XXI. p. 319). Of these, after omitting 8 and 9, there remain 568; the diagram represents the course traced out by following the direction of these as the clue to our path. Many of the steps have of course been taken in opposite directions twice or oftener. The result seems to me to furnish a very fair graphical indication of randomness. I have compared it with corresponding paths furnished by rows of figures taken from logarithmic tables, and in other ways, and find the results to be much the same.

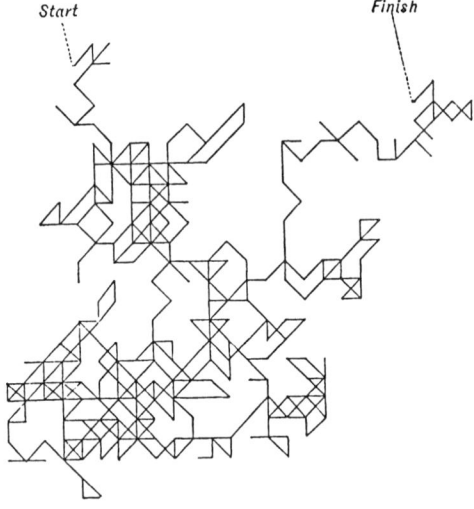

Start   *Finish*

Estimating a Binomial Parameter Using the Likelihood Function

Comments on Thiele (1889)

1 Introduction

As was the case with probability in the seventeenth century, likelihood had been emerging as a new concept for some time before it was formalized and named (by Fisher, 1921). Early examples of the method of maximum likelihood have been noticed in the work of J.H. Lambert and Daniel Bernoulli in the eighteenth century and discussed by several authors (see Edwards, 1974; Hald, 1998, Chapter 5; and Stigler, 1999, Chapter 16, for introductions to the literature), and the emergence of the method in modern times has been discussed by Aldrich (1997), Edwards (1997b), and Hald (1998, Chapter 28; 1999).

The use of likelihood itself for direct statistical inference is distinct from the use of the method of maximum likelihood as a method of point estimation. Although we might somewhat anachronistically now be inclined to interpret the eighteenth-century "argument from design" in terms of the ratio of likelihoods rather than of probabilities (see our Comments on Arbuthnott, 1710), the suggestion that the likelihood could itself be interpreted as a relative measure of support had seemed first to appear in Fisher (1922) and in particular in a footnote in which Fisher remarked, "Although, in an important class of cases, the likelihood may be held to measure the degree of our rational belief in a conclusion . . ." His final endorsement of the idea that, in appropriate cases, "The likelihood supplies a natural order of preference among the possibilities under consideration" is to be found in Fisher (1956), Chapter 3, complete with a diagram of the likelihood function for a parameter. These and other relevant quotations from Fisher may be found in Edwards (1972, 1992, Chapters 2 and 3), together with an extended account of the likelihood approach to inference.

2 Thiele's *Lectures on the General Theory of Observations: Calculus of Probability and the Method of Least Squares* (1889)

Until now the only earlier hint of the direct use of likelihood was thought to have been a comment by Fisher himself in his very first paper (1912;

reprinted in Edwards, 1997a) in which he advocated the use of the method of maximum likelihood (though not under that name). He there wrote that the quantity "P," which he was later to call the likelihood, "is a relative probability only, suitable to compare point with point...." However, Professor Steffen Lauritzen has pointed out to us that Thiele (1889) had already invoked the concept of likelihood (*"rimelighed"*) in his discussion of the problem of the estimation of a binomial parameter (see Lauritzen, 1999). As with Fisher's initial comments, Thiele is tentative in his suggestions, but his account is accompanied by a diagram of considerable interest and originality in which likelihoods are graphed. It is indeed interesting to note the similarity in the ways in which Thiele and Fisher express the fact that likelihoods are not to be manipulated like probabilities.

It is almost certain that Fisher was not familiar with Thiele's book. (Hald, 1998, remarks, "Fisher was a genius who almost single-handedly created the foundations for modern statistical science without detailed study of his predecessors. When young he was ignorant not only of the Continental contributions but even of contemporary contributions in English.") In his discussions of Thiele's work Hald (1981, 1998, Chapter 27) does not comment on the use of likelihood, and Thiele himself did not discuss it in the more popular version of his book which was translated into English in 1903 and later reprinted in the *Annals of Mathematical Statistics* (Thiele, 1931).

Section VIe of Thiele (1889) is headed *Indirect or a posteriori treatment of probability from observed relative frequency.* It is somewhat discursive and we summarize much of it. He discusses the point estimation of the parameter $p/(p + q)$ from m successes and l failures in n trials, and we immediately put $p + q = 1$ without loss of generality in order to simplify the notation. Thiele first observes that no values of p except 0 and 1 are completely impossible (he seems to ignore the cases of either m or l being zero), and he then points out that any estimator $(m+a)/(m+l+b)$, where a and b are arbitrary positive constants with b greater than a, is acceptable because it has the property of consistency. He does not, of course, use the modern terminology, but his meaning is clear. Next he requires $b = 2a$ on the grounds of symmetry, since estimating q should lead to the same conclusion as estimating $1 - p$. What number to choose for a "depends on my attitude towards the consequences of such a choice."

In the literature on this question, observes Thiele, only values of a between, and including, 0 and 1 have been seriously entertained, and to investigate the choice further we must study the error law for what was observed and what was not observed but might have been, in other words, the binomial distribution with parameter p for n trials. He gives the distribution function and its first three "half-invariants," and recommends the reader draw some examples, from which he will learn that the distribution is unimodal but not in general symmetrical. Thiele gives details of the mode, remarking that "a more definite value for the maximum of the [continuous] error curve can only be given under a somewhat arbitrary assumption" about interpolation between the integer values of m.

3   The Probability/Likelihood Diagram

Thiele then explains what he has in mind by means of a figure (Figure 1) for the case $n = 10$ "under the assumption that the continuous error curve can be determined by ordinary interpolation using Newton's or similar methods." We do not quote his detailed description, since the figure is immediately clear to the modern reader. It is a contour diagram of the binomial probabilities (interpolated as required) with m as the x-axis and p as the y-axis. For each value of p the profile of a transect is the corresponding binomial distribution (with interpolation) for m, and for each value of m the profile of a transect is the corresponding likelihood function (as we should now say) for p. "$n = 10$ has been chosen because on the one hand it accentuates the symmetry around the diagonal which becomes the most apparent feature of the curves as n increases, whilst on the other hand $n = 10$ is still small enough to display a characteristic skewness, breaking this symmetry." By coincidence, Figure 3 of Edwards (1972, 1992) also displays the binomial likelihood curves for $n = 10$, $m = 1$ to 9, which are therefore profiles of Thiele's figure. Thiele comments on the main geometrical features of his figure, which extends into the irrelevant regions $m < 0$ and $m > 10$.

Thiele now returns to the question of the value of a. He has added to his figure the lines $p = (m + a)/(m + l + 2a)$ for $a = 0$, $\frac{1}{2}$, and 1. "The most central of these straight lines ($a = \frac{1}{2}$) intersects the contours of the figure at least so close to the points of maxima [in the x-direction] that drawings like this do not show the proper difference," and "considering the undetermined definition of the continuous error curves … one can very well think of the straight line for $a = \frac{1}{2}$ as the geometric locus of the maxima." After remarking that the value $a = 0$ corresponds to equating the observed and expected means, Thiele continues with the extract which we print.

The extract concludes with a reference to F. Bing's celebrated objection to Bayesian inference, that the choice of a prior distribution to represent ignorance depends on the parametrization of the model (see Hald, 1998, Chapter 15), but Thiele goes on to remark "Yet I do not think that for this reason one should discard the method completely," and if one assumes that the likelihoods can be treated as probabilities one would recover Bayes's rule.

Most of the rest of Thiele's discussion is devoted to a consideration of the repeated-sampling consequences of adopting particular values of a, for "Whatever value one chooses, it should not be considered an exact number, but only the main value in an error law for the probability [p]. Any such determination needs to be supplemented with an indication of additional details for the error law concerning the likely limits for the result." In the end, he inclines to the value $a = \frac{1}{2}$.

4   Summary

In his Section VIe Thiele has presented a novel analysis of the problems associated with inferring the value of a binomial parameter, introducing the idea of a consistent estimator and that of the likelihood function for a parameter. He discusses the maximum likelihood estimator, the Bayesian estimator, and the repeated-sampling characteristics of his class of estimators in general, balancing the advantages and disadvantages of the several viewpoints. It may be noted that Thiele did not apply his likelihood approach elsewhere in his book, when considering least squares, for example.

5   A Note on the "Continuous" Binomial Distribution

We conclude with some remarks on the problem of interpolating the binomial distribution so as to derive a continuous error curve, as outlined by Thiele in 1889, but not then pursued by him beyond the suggestion of using "Newton's formula or similar theories."

The natural way to "interpolate" the binomial coefficients in order to produce a continuous curve is simply to replace the factorials such as $n!$ in the coefficients by the gamma function $\Gamma(n+1)$ and express the result in terms of the B-function:

$$\binom{n}{m} = 1/((n+1)B(m+1, nm+1)) \ .$$

Interpolation as such is then unnecessary, and the resulting continuous curve is positive in the relevant range $0 \le m \le n$ and passes through the required values for integral m. Moreover, it possesses the "additive" property of binomial coefficients extended to the continuous case, as may easily be proved. Thiele did not suggest this curve, but in his later book (originally in Danish, 1897; English reprint Thiele, 1931, p. 185) he suggested the formula

$$\frac{n!}{(n-m)(n-1-m)\ldots(1-m)} \frac{\sin \pi m}{\pi m} \ .$$

This is undefined for integral m but in other respects is the identical function. However, he rejected it because outside the range it included negative values, and it seems that he was seeking an error curve on the whole real line and not just on part of it.

Once a smooth curve incorporating the binomial coefficients has been chosen the continuous binomial distribution follows immediately, for any value of its parameter p.

References

Aldrich, J. (1997). R.A. Fisher and the making of maximum likelihood 1912–1922. *Statist. Sci.*, **12**, 162–176.

Edwards, A.W.F. (1972, 1992). *Likelihood.* Cambridge University Press. Expanded Edition 1992, Johns Hopkins University Press, Baltimore.)

Edwards, A.W.F. (1974). The history of likelihood. *Intern. Statist. Rev.*, **42**, 9–15. Reprinted in *Likelihood*, Expanded Edition 1992, Johns Hopkins University Press, Baltimore.

Edwards, A.W.F. (1997a). Three early papers on efficient parametric estimation. *Statist. Sci.*, **12**, 35–47.

Edwards, A.W.F. (1997b). What did Fisher mean by "inverse probability" in 1912–22? *Statist. Sci.*, **12**, 177–184.

Fisher, R.A. (1912). On an absolute criterion for fitting frequency curves. *Messenger. Math.*, **41**, 155–160.

Fisher, R.A. (1921). On the "probable error" of a coefficient of correlation deduced from a small sample. *Metron*, **1**, 3–32.

Fisher, R.A. (1922). On the mathematical foundations of theoretical statistics. *Phil. Trans. Roy. Soc.*, A, **222**, 309–368.

Fisher, R.A. (1956). *Statistical Methods and Scientific Inference.* Oliver & Boyd, Edinburgh.

Hald, A. (1981). T.N. Thiele's contributions to statistics. *Intern. Statist. Rev.*, **49**, 1–20.

Hald, A. (1998). *A History of Mathematical Statistics from 1750 to 1930.* Wiley, New York.

Hald, A. (1999). On the history of maximum likelihood in relation to inverse probability and least squares. *Statist. Sci.*, **14**, 214–222.

Lauritzen, S. L. (1999). Aspects of T. N. Thiele's contributions to statistics. *Bull. Int. Statist. Inst.* **58**, 27–30.

Stigler, S.M. (1999). *Statistics on the Table.* Harvard University Press, Cambridge, MA.

Thiele, T.N. (1889). *The General Theory of Observations: Calculus of Probability and the Method of Least Squares* (in Danish). Reitzel, Copenhagen.

Thiele, T.N. (1931). The theory of observations. *Ann. Math. Statist.*, **2**, 165–307.

Thiele (1889), Section VIe (extract)

Note: *Thiele refers throughout to the binomial parameter [p] as "the probability" because his context is that of estimating the probability of success in a future trial, but this leads to occasional lack of clarity. We have therefore added "[p]" when it is the parameter that is meant. By "Bayes's rule" Thiele means what is often called "Laplace's rule of succession," namely the value for the probability [p] which arises when Bayes's uniform prior is adopted for it.*

Thus the values 0 and $\frac{1}{2}$ for a commend themselves by directing the choice of the probability in question [p] to the particularly interesting values for which the observed frequency is equal to the *mean* or center of gravity of the error curve in the first case $(a = 0)$ and to its *maximal value* in the second case $(a = \frac{1}{2})$. The choice $a = 1$ does not commend itself through any such simple and conspicuous property, but although this choice does not lead to the probability [p] which makes the observed outcome the most probable of all and, as our figure shows, gives only a slightly greater probability than that which corresponds to $a = 0$, it does nevertheless lead to probabilities which make the observed outcome relatively highly probable.

The assumption $a = 1$, which under the name of Bayes's rule has played an important role and at least has historical significance, has hardly any more distinctive property related to these error curves and their representation in figures such as ours for $m + l = 10$ than that which is associated with the contour-figure's transects along the straight lines for the observed frequencies; such transects, normal to those previously mentioned, would also represent lines with properties characteristic of skew error curves, and whereas the maximal heights in these transects now correspond to $a = 0$, the center of gravity of the curves, or the mean probabilities, are determined by their intersections with $a = 1$.

[Thiele then derives the mode and mean of these curves, finding them to be $m/(m + l)$ and $(m + 1)/(m + l + 2)$, respectively. We omit the next paragraph, which outlines a possible generalization of the binomial model.]

A similar consideration lies at the bottom of the previously accepted theory of "a posteriori probabilities." When a single relative frequency $[m/(m + l)]$ is observed and one is uncertain about which of the infinitely many possible probabilities [p] between 0 and 1 to derive from this frequency, then the thought occurs that although all probabilities are possible, they will evidently not all be equally likely [*rimelige*] as hypothetical causes of the observed frequency; the likelihood [*rimelighed*] must for each of them be expressible in terms of a number, and it is not obviously incorrect to measure this likelihood with the probability $\phi(m/(m + l))$ [the binomial distribution function] of the observed frequency, calculated on the assumption of p being the unknown value for the probability. That for these measures of the likelihood of the hypotheses people have used the term "the probability of the correctness of the individual hypotheses" would not have done any harm unless they had additionally allowed themselves to calculate with these "probabilities" using all the rules from the direct calculus of probability, even though the definition of probability does not apply to them. In *Tidsskrift for Mathematik*, 1879, Director Bing has however shown that such an extension of the notion of probability leads to contradictions when applied without additional precautions.

Yet I do not think that for this reason one should discard the method completely, in particular if it is only applied to cases where the observation is either/or. If one does assume that the likelihoods for the individual hypothetical probabilities [p] in this case can be treated as probabilities or according to the rules of mathematical expectation and are considered proportional to the probability of the observed frequency as given by each hypothesis, then it is proper that the required probability be determined by Bayes's rule ... $(m + 1)/(m + l + 2)$.

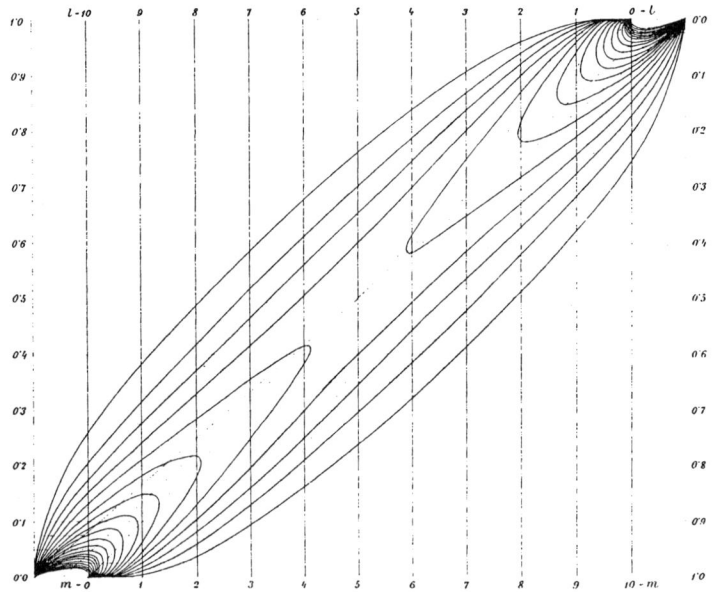

FIGURE 1. Thiele's figure described in the comments and the extract.

Yule's Paradox ("Simpson's Paradox")

Comments on Yule (1903)

George Udny Yule may be described as the father of the contingency table by virtue of his long pioneering paper, "On the association of attributes in statistics" published in the *Philosophical Transactions of the Royal Society* in 1900. At the time Yule was closely associated with Karl Pearson at University College London, himself then deeply involved in developing coefficients of correlation and other measures of association in 2×2 tables arising from the grouping of normally distributed variables (Pearson introduced the term "contingency table" in 1904). A dozen years later the differing approaches of the two men led to an acrimonious public controversy (described by Mackenzie, 1981; see also Magnello, 1993, and Aldrich, 1995) which, however, lies outside our present topic, the creation of spurious associations through combining heterogeneous material.

In 1899 Pearson had entitled a section of one of his *Philosophical Transactions* papers on evolution "On the spurious correlation produced by forming a mixture of heterogeneous but uncorrelated material" (Pearson, 1899, Section 7 of Part I). For example, a population consisting of two sub-populations described by bivariate normal distributions with zero correlation parameters but different means in respect of both variates will obviously exhibit a correlation. Then in 1903 Yule extended his 1900 work by considering the nature of complete independence and the analogue of Pearson's "spurious correlation" for a contingency table. It is this section of his paper which we reproduce, "On the fallacies that may be caused by the mixing of distinct records," in which Yule gives an artificial numerical example showing how it is possible to have no association in each of two 2×2 tables but an association when the two are combined.

In his textbook *An Introduction to the Theory of Statistics*, which ran to ten editions by 1932, Yule (1911) treated "partial association" in Chapter 4, extending the discussion to cases of arbitrary association. It is reasonable, therefore, to describe as "Yule's paradox" the phenomenon that the association between A and B in a combined class $\{C$ and not$-C\}$ does not generally depend only on the associations in $\{C\}$ and $\{$not$-C\}$ separately, because of the possible influence of associations between A and C and between B and C. Exercise 6 of Chapter 4 of Yule's book is intended to illustrate a case of negative associations in both $\{C\}$ and $\{$not$-C\}$ and a positive association in $\{C$ and not$-C\}$, though it seems that some of the data have been inadvertently omitted. We make the same point with another example (Edwards, 1963):

Male				Female		
	Untreated	Treated			Untreated	Treated
Alive	3	5		Alive	9	6
Dead	4	6		Dead	5	3

The treatment is equally beneficial to both males and females (as measured by the odds ratios) but detrimental in the combined sample.

The section was repeated in the eleventh edition written jointly with M.G. Kendall; even the incomplete data survived (Yule and Kendall, 1937). When, in turn, Kendall wrote his *The Advanced Theory of Statistics* he dealt with "illusory association" in like manner (see, for example, the second edition, Kendall, 1945). These very successful books were the principal texts available in Britain in their time, yet the next generation of textbooks for the most part ignored the problem (an exception was Bailey, 1959). When Simpson (1951) wrote his oft-quoted paper, "The interpretation of interaction in contingency tables," he assumed familiarity with Yule's paradox (then unnamed): "The dangers of amalgamating 2×2 tables are well known and are referred to, for example, on p. 317 of Kendall (1945), Vol. I," and he did not refer to Yule's papers.

Simpson's paper led to an increased interest in Yule's paradox, but not one of the flurry of papers in the 1960s referred to Yule's papers either (Darroch, 1962; Lewis, 1962; Plackett, 1962; Birch, 1963; Good, 1963; Edwards, 1963; Goodman, 1964), though Lewis, like Simpson, did mention Yule himself: "The fact that a 2×2 table showing association can be decomposed into two or more tables showing no association has been well known since Yule's classic discussion on 'illusory association' ... (Yule and Kendall, 1950)." Important earlier and later papers on the topic were also silent (Roy and Kastenbaum, 1956, Ku and Kullback, 1968). Most authors simply assumed that Bartlett's (1935) short note suggesting a criterion for no second-order interaction in $2 \times 2 \times 2$ table was the starting point, and it happened that Bartlett did not refer to the earlier literature. In a fine example of Stigler's Law ("No law is named after its true originator," Stigler, 1980), Yule's paradox thus became "Simpson's paradox" (see, for example, Blyth, 1972, "On Simpson's paradox and the sure-thing principle," and the book by Cox and Snell, 1989, *Analysis of Binary Data*, neither of which refer to Yule).

In the extract from Yule (1903) which follows, (A), (B), and (C) are the numbers possessing each of three attributes A, B, and C, and (α), (β), and (γ) the numbers not possessing them, the total number $(A) + (\alpha)$, etc., being given by N. Furthermore, $(A\gamma)$, for example, is the number possessing A but not C, $(A\beta C)$ the number possessing A and C but not

B, and so on. Finally, $|AB|$, for example, means the association between A and B (as measured by Yule's well-known coefficient Q; Yule, 1900) and $|AB|C|$ the association between A and B amongst the Cs.

References

Aldrich, J. (1995). Correlations genuine and spurious in Pearson and Yule. *Statist. Sci.*, **10**, 364–376.

Bailey, N.T.J. (1959). *Statistical Methods in Biology*. English Universities Press, London.

Bartlett, M.S. (1935). Contingency table interactions. *J. Roy. Statist. Soc.*, Suppl. **2**, 248–252.

Birch, M.W. (1963). Maximum likelihood in three-way contingency tables. *J. Roy. Statist. Soc.* B, **25**, 220–233.

Blyth, C.R. (1972). On Simpson's paradox and the sure-thing principle. *J. Amer. Statist. Assn.*, **67**, 364–366.

Cox, D.R. and Snell, E.J. (1989). *Analysis of Binary Data*, 2nd edn. Chapman and Hall, London.

Darroch, J.N. (1962). Interactions in multi-factor contingency tables. *J. Roy. Statist. Soc.* B, **24**, 251–263.

Edwards, A.W.F. (1963). The measure of association in a 2×2 table. *J. Roy. Statist. Soc.* A, **126**, 109–114.

Good, I.J. (1963). Maximum entropy for hypothesis formulation, especially for multidimensional contingency tables. *Ann. Math. Statist.*, **34**, 911–934.

Goodman, L.A. (1964). Interactions in multidimensional contingency tables. *Ann. Math. Statist.*, **35**, 632–646.

Kendall, M.G. (1945). *The Advanced Theory of Statistics*, Vol. 1, 2nd edn. Griffin, London.

Ku, H.H. and Kullback, S. (1968). Interaction in multidimensional contingency tables: An information theoretic approach. *J. Research Nat. Bureau Standards—Math. Sci.*, **72B**, 159–199.

Lewis, B.N. (1962). On the analysis of interaction in multi-dimensional contingency tables. *J. Roy. Statist. Soc.* A, **125**, 88–117.

Mackenzie, D.A. (1981). *Statistics in Britain 1865–1930*. Edinburgh University Press.

Magnello, M.E. (1993). Karl Pearson: Evolutionary biology and the emergence of a modern theory of statistics (1884–1936). D.Phil. Thesis, University of Oxford.

Pearson, K. (1899). Mathematical contributions to the theory of evolution. VI. Part I. *Phil. Trans. Roy. Soc.* A, **192**, 257–330.

Pearson, K. (1904). Mathematical contributions to the theory of evolution. XIII. On the theory of contingency and its relation to association and normal correlation. *Drapers' Company Research Memoirs*, Biometric Ser. I.

Plackett, R.L. (1962). A note on interaction in contingency tables. *J. Roy. Statist. Soc. B*, **24**, 162–166.

Roy, S.N. and Kastenbaum, M.A. (1956). On the hypothesis of no "interaction" in a multi-way contingency table. *Ann. Math. Statist.*, **27**, 749–757.

Simpson, E.H. (1951). The interpretation of interaction in contingency tables. *J. Roy. Statist. Soc. B*, **13**, 238–241.

Stigler, S.M. (1980). Stigler's law of eponymy. *Trans. New York Acad. Sci. Ser. II*, **39**, 147–157.

Yule, G.U. (1900). On the association of attributes in statistics: With illustrations from the material of the Childhood Society, &c. *Phil. Trans. Roy. Soc. A*, **194**, 257–319.

Yule, G.U. (1903). Notes on the theory of association of attributes in statistics. *Biometrika*, **2**, 121–134.

Yule, G.U. (1911). *An Introduction to the Theory of Statistics*. Griffin, London.

Yule, G.U. and Kendall, M.G. (1937). *An Introduction to the Theory of Statistics*, 11th edn. Griffin, London.

Yule, G.U. and Kendall, M.G. (1950). *An Introduction to the Theory of Statistics*, 14th edn. Griffin, London.

On the Theory of Association

5. *On the fallacies that may be caused by the mixing of distinct records.*

It follows from the preceding work that we cannot infer independence of a pair of attributes within a sub-universe from the fact of independence within the universe at large. From $|AB| = 0$, we cannot infer $|AB|C| = 0$ or $|AB|\gamma| = 0$, although we can of course make the corresponding inference in the case of complete association—i.e. from $|AB| = 1$ we do infer $|AB|C| = |AB|\gamma| = \text{etc.} = 1$. But the converse theorem is also true; a pair of attributes does not necessarily exhibit independence within the universe at large even if it exhibit independence in *every* sub-universe; given

$$|AB|C| = 0, \qquad |AB|\gamma| = 0,$$

we cannot infer $|AB| = 0$. The theorem is of considerable practical importance from its inverse application; i.e. even if $|AB|$ have a sensible positive or negative value we cannot be sure that nevertheless $|AB|C|$ and $|AB|\gamma|$ are not both zero. Some given attribute might, for instance, be inherited neither in the male line nor the female line; yet a mixed record might exhibit a considerable apparent inheritance. Suppose for instance that 50 $°/_0$ of the fathers and of the sons exhibit the attribute, but only 10 $°/_0$ of the mothers and daughters. Then if there be no inheritance in either line of descent the record must give (approximately)

> fathers with attribute and sons with attribute 25 $°/_0$
>
> „   „   „   „   without   „   25 $°/_0$
>
> „   without   „   „   with   „   25 $°/_0$
>
> „   „   „   „   without   „   25 $°/_0$
>
> mothers with attribute and daughters with attribute   1 $°/_0$
>
> ,   „   „   „   „   without   „   9 $°/_0$
>
> „   without   „   „   „   with   „   9 $°/_0$
>
> „   „   „   „   „   without   „   81 $°/_0$.

If these two records be mixed in equal proportions we get

> parents with attribute and offspring with attribute 13 $°/_0$
>
> „   „   „   „   „   without   „   17 $°/_0$
>
> „   without   „   „   „   with   „   17 $°/_0$
>
> „   „   „   „   „   without   „   53 $°/_0$

Here $13/30 = 43\frac{1}{3}$ $°/_0$ of the offspring of parents with the attribute possess the attribute themselves, but only 30 $°/_0$ of offspring in general, i.e. there is quite a large but illusory inheritance created simply by the mixture of the two distinct records. A similar illusory association, that is to say an association to which the most obvious physical meaning must not be assigned, may very probably occur in any other case in which different records are pooled together or in which only one record is made of a lot of heterogeneous material.

Consider the case quite generally. Given that $|AB|C|$ and $|AB|\gamma|$ are both zero, find the value of (AB). From the data we have at once

$$(AB\gamma) = \frac{(A\gamma)(B\gamma)}{(\gamma)} = \frac{[(A)-(AC)][(B)-(BC)]}{[N-(C)]},$$

$$(ABC) = \frac{(AC)(BC)}{(C)}.$$

Adding

$$(AB) = \frac{N(AC)(BC)-(A)(C)(BC)-(B)(C)(AC)+(A)(B)(C)}{(C)[N-(C)]}.$$

Write

$$(AB)_0 = \frac{1}{N}(A)(B), \quad (AC)_0 = \frac{1}{N}(A)(C), \quad (BC)_0 = \frac{1}{N}(B)(C),$$

subtract $(AB)_0$ from both sides of the above equation, simplify, and we have

$$(AB) - (AB)_0 = \frac{N[(AC) - (AC)_0][(BC) - (BC)_0]}{C[N - (C)]}.$$

That is to say, *there will be apparent association between A and B in the universe at large unless either A or B is independent of C.* Thus, in the imaginary case of inheritance given above, if A and B stand for the presence of the attribute in the parents and the offspring respectively, and C for the male sex, we find a positive association between A and B in the universe at large (the pooled results) because A and B are both positively associated with C, i.e. the males of both generations possess the attribute more frequently than the females. The " parents with attribute" are mostly males ; as we have only noted offspring of the same sex as the parents, their offspring must be mostly males in the same proportion, and therefore more liable to the attribute than the mostly-female offspring of "parents without attribute." It follows obviously that if we had found no inheritance to exist in any one of the *four* possible lines of descent (male-male, male-female, female-male, and female-female), no fictitious inheritance could have been introduced by the pooling of the *four* records. The pooling of the two records for the crossed-sex lines would give rise to a fictitious negative inheritance—disinheritance—cancelling the positive inheritance created by the pooling of the records for the same-sex lines. I leave it to the reader to verify these statements by following out the arithmetical example just given should he so desire.

The fallacy might lead to seriously misleading results in several cases where mixtures of the two sexes occur. Suppose for instance experiments were being made with some new antitoxin on patients of both sexes. There would nearly always be a difference between the case-rates of mortality for the two. If the female cases terminated fatally with the greater frequency and the antitoxin were administered most often to the males, a *fictitious* association between " antitoxin " and " cure" would be created at once. The general expression for $(AB) - (AB)_0$ shews how it may be avoided; it is only necessary to *administer the antitoxin to the same proportion of patients of both sexes.* This should be kept constantly in mind as an essential rule in such experiments if it is desired to make the most use of the results.

The fictitious association caused by mixing records finds its counterpart in the spurious correlation to which the same process may give rise in the case of continuous variables, a case to which attention was drawn and which was fully discussed by Professor Pearson in a recent memoir[*]. If two separate records, for each of which the correlation is zero, be pooled together, a spurious correlation will necessarily be created unless the mean of one of the variables, at least, be the same in the two cases.

[*] *Phil. Trans.* A, Vol. 192, p. 277.

Beginnings of
Extreme-Value Theory

Comments on Bortkiewicz (1922a) and von Mises (1923)

1 Introduction

The distribution of the largest or the smallest of n iid variates naturally has a very long history and goes back at least to Nicholas Bernoulli in 1709. Bernoulli reduces a problem of the expected lifetime of the last survivor among n men to finding the expected value of the maximum of n iid uniform variates. Harter (1978) summarizes this and numerous other early papers that touch on the extremes and the range. Gumbel (1958) gives a brief historical account.

An intensive development of extreme-value theory seems to have been stimulated by two overlapping papers by Bortkiewicz (1922a,b), dealing not only with the range but also with the ordered *absolute* errors $\epsilon_1 \leq \epsilon_2 \leq \cdots \leq \epsilon_n$. A symmetrical parent distribution, generally the normal, is assumed. The (1922b) paper, an expanded but less focused version of (1922a), contains a review of earlier work, especially on the maximum absolute error ϵ_n.

Here we present, in translations from the German, the early portion of Bortkiewicz (1922a) and the paper it triggered, the pioneering article by von Mises (1923), where some truly asymptotic results are rigorously obtained.

To a surprising extent both authors still use Gauss's terms and symbols. We have (a) replaced their "mean error" (mittlerer Fehler) by "standard deviation" and the corresponding symbol μ by σ, (b) retained the precision $h = 1/\sigma\sqrt{2}$, (c) *returned* to Gauss's Θ-function to denote the error function

$$\Theta(x) = \frac{2}{\sqrt{\pi}} \int_o^x e^{-t^2} dt,$$

and set $\theta(x) = \Theta'(x)$. The authors confusingly write $\Phi(x)$, $\phi(x)$ for our $\Theta(x), \theta(x)$. Note that, with $\Phi(x)$ and $\phi(x)$ having their *modern* meaning,

$$\Theta(x) = 2\Phi(\sqrt{2}x) - 1, \ \theta(x) = 2\sqrt{2}\phi(\sqrt{2}x).$$

2   Finite-Sample Results

Bortkiewicz's derivation of his interesting relation (1), valid for any symmetric parent distribution, is ingenious. However, it is far from being the most convenient expression for $\bar{v}$, the expected value of the range. Von Mises notes that for *any* parent distribution with cdf $W(x)$ one has $\bar{v} = g_n - k_n$, the difference between the expectations of the largest and the smallest observations. This enables him to arrive quickly at his result (3). His use of Stieltjes integrals to cover also discrete parent distributions seems original in the context of order statistics. Nevertheless, von Mises missed obtaining the better expression

$$\bar{v} = n \int_{-\infty}^{\infty} [1 - W^n(x) - (1 - W(x))^n] dx, \qquad (a)$$

which follows on integration by parts of his eq. (3).

As one has come to expect, both authors seem unaware of Pearson's (1902) result for the expectation of the difference of successive order statistics:

$$E(X_{(r+1)} - X_{(r)}) = \binom{n}{r} \int_{-\infty}^{\infty} W^r(x)[1 - W(x)]^{n-r} dx. \qquad (b)$$

As noted by Tippett (1925), eq. (a) above follows by summing (b) from $r = 1$ to $r = n$. However, Tippett assumes an absolutely continuous parent distribution. A formal justification of (a) in the discrete case had to wait for Cox (1954). See also David (1981, p. 38).

3   Asymptotic Theory

In the latter part of Bortkiewicz (1922a), not translated here, approximations to $\bar{v}$ are examined that are better than $\bar{v} \doteq 2M_{n/2}$, with which we end. However, these approximations are rather ad hoc and no attempt is made to handle the limiting case $n \to \infty$.

Von Mises, in his Section 2, gives a careful proof that $\lim_{n\to\infty} (g_n/x_n) = 1$, where $x_n = W^{-1}(1 - \frac{1}{n})$, the cdf $W(x)$ being here assumed to be a monotone increasing function satisfying, for fixed ξ,

$$\lim_{x\to\infty} \frac{1 - W(x + \xi)}{1 - W(x)} = 0. \qquad (c)$$

Condition (c) is typical of the tail-behavior assumptions made in subsequent extreme-value theory work. The quantity x_n has become standard as an asymptotic measure of location of the largest value, $X_{(n)}$.

The final paragraph of Section 2, following the proof of the main theorem, needs commentary. Let us consider n Bernoulli trials involving the n iid variates $X_1, \ldots, X_n$, a "success" occurring when $X_i > W^{-1}(1 - \frac{1}{m})$, where m is fixed. If S_n denotes the total number of successes, then, by the strong law of large numbers, with probability 1 only finitely many of the events

$$\left| \frac{S_n}{n} - \frac{1}{m} \right| > \epsilon$$

occur for any $\epsilon > 0$. In other words, S_n/n converges almost surely to $1/m$, so that the proportion of observations to the left of x_m converges almost surely to $1 - 1/m$.

However, von Mises's tentatively worded extension from fixed m to $m = n$ is incorrect since

$$\lim_{n \to \infty} P(S_n = 0) = \lim_{n \to \infty} \left(1 - \frac{1}{n} \right)^n = e^{-1} = 0.36788,$$

so that the largest observation does not almost surely lie to the right of x_n.

This aspect is better handled by Dodd (1923) who is the first to obtain *explicit* results for parent distributions other than the normal. Of course, von Mises's theorem above is applicable to a large class of distributions and not just to the normal, as has been claimed by several writers.

Typical of Dodd's results is the following theorem. "If the probability function is

$$\phi(x) = g^{x^\alpha} \cdot \psi(x), \text{ with } x^{-\beta} < \psi(x) < x^\beta,$$

where α, β, g are positive constants, and $g < 1$, then it is asymptotically certain that the greatest of n variates will equal

$$(-\log_g n)^{1/\alpha}(1 + \epsilon'), \text{ with } |\epsilon'| < \epsilon,$$

small at pleasure."

If we set

$$g = e^{\frac{1}{2\sigma^2}}, \ \alpha = 2, \ \beta = 0, \ \psi(x) = 1,$$

then

$$(-\log_g n)^{1/\alpha} = \sigma \sqrt{2 \log_e n},$$

which is asymptotically equivalent to z_o in von Mises's eq. (17), for which $\sigma = 1/\sqrt{2}$. (Note that in the reproduction of Dodd's summary table in Harter (1978, p. 158) $-\log_g n$ has been misprinted as $-\log g^n$.)

Subsequent important early papers on extreme-value theory, not necessarily citing Bortkiewicz or von Mises, include Tippett (1925), Fréchet (1927), Fisher and Tippett (1928), Gumbel (1935), von Mises (1936), and Gnedenko (1943). Apart from Gumbel (1958) and Harter (1978), see the modern account by Galambos (1987). The impressive paper by Gnedenko has been translated by N.L. Johnson in Kotz and Johnson (1992) and is discussed by R.L. Smith.

Briefly stated, von Mises (1936) provides useful necessary conditions on the parent distribution for the appropriately normalized maximum to approach, with increasing n, one of the three limiting distributions shown to be the only ones possible by Fisher and Tippett (1928). Gnedenko (1943) goes on to give necessary and sufficient conditions.

Some other areas of von Mises's research are reviewed by Cramér (1953), especially his writings on the foundations of probability. Although von Mises's extensive attempt to define probability as the limit of a relative frequency lost out to the set-theoretical approach, he was important in the clarification of ideas. His work on statistical functions remains important in nonparametric theory.

Von Mises appears to have introduced, in German, two basic statistical terms whose literal translations into English are now standard. The terms are "distribution function" (Verteilungsfunktion) and "probability density" (Wahrscheinlichkeitsdichte). Both occur in his paper on the foundations of probability theory (von Mises, 1919).

4   A Biographical Note on Richard von Mises

We conclude with a few remarks on von Mises, since, surprisingly, no biographical sketch on him appears in Johnson and Kotz (1997). However, there is an entry in the *Encyclopedia of Statistical Sciences*. A selection of his papers has been published in two volumes (von Mises, 1964). The first volume, *Geometry, Mechanics, Analysis*, includes a biographical note by S. Goldstein. Volume 2, *Probability and Statistics, General*, lists his many publications, including 14 books.

The extraordinarily versatile von Mises was born in 1883 in Lemberg, in the Austro-Hungarian Monarchy, now Lvov, Ukraine. He studied mechanical engineering at the Vienna Technical University. His first appointment, in Applied Mathematics at the University of Strassburg, came to an end with the first world war. Von Mises volunteered for the Austro-Hungarian army, where he was a flight instructor and also designed and supervised the construction of the first large airplane of the Monarchy. In 1920 he was appointed Professor and Director of Applied Mathematics at the University of Berlin.

During his 13 years in Berlin, von Mises was engaged not only in applied mathematics, but also in probability and philosophy. His most popular book *Wahrscheinlichkeit, Statistik und Wahrheit* was first published in 1928. In 1939 its second edition was translated into English (*Probability, Statistics and Truth*) by J. Neyman et al. Also in Berlin von Mises began seriously to build the greatest private collection in the world of the works of the German poet Rilke, on whom he became an authority.

In 1921 von Mises founded the *Zeitschrift für angewandte Mathematik und Mechanik*. This journal is still prominent and since 1984 carries the additional title *Applied Mathematics and Mechanics*, listing him as founding editor. Von Mises's term as editor ended abruptly in 1933 with Hitler's ascent to power. He was Jewish only on Nazi criteria, having converted to Catholicism during his Strassburg years. Although as a veteran of World War I he would have been allowed to stay on for another two years, he accepted an invitation to Istanbul, as Professor of Mathematics and Director of the Mathematical Institute. There he frequently published in French, switching to English upon joining Harvard University in 1939. He soon became Gordon McKay Professor of Aerodynamics and Applied Mathematics. Von Mises died in Boston in 1953.

References

Bortkiewicz, L. von (1922a). Variationsbreite und mittlerer Fehler. *Sitzungsberichte der Berliner Math. Gesellschaft*, **21**, 3–11.

Bortkiewicz, L. von (1922b). Die Variationsbreite beim Gaussschen Fehlergesetz. *Nordisk Statistisk Tidskrift*, **1**, 11–38, 193–220.

Cox, D.R. (1954). The mean and coefficient of variation of range in small samples from non-normal populations. *Biometrika*, **41**, 469–481.

Cramér, H. (1953). Richard von Mises' work in probability and statistics. *Ann. Math. Statist.*, **24**, 657–662.

David, H.A. (1981). *Order Statistics*, 2nd edn. Wiley, New York.

Dodd, E.L. (1923). The greatest and the least variate under general laws of error. *Trans. Amer. Math. Soc.*, **25**, 525–539.

Fisher, R.A. and Tippett, L.H.C. (1928). Limiting forms of the frequency distribution of the largest or smallest member of a sample. *Proc. Camb. Phil. Soc.*, **24**, 180–190.

Fréchet, M. (1927). Sur la loi de probabilité de l'écart maximum. *Ann. Soc. Polonaise de Math.* (Cracow), **6**, 93–116.

Galambos, J. (1987). *The Asymptotic Theory of Extreme Order Statistics*, 2nd edn. Krieger, Malabar, FL.

Gnedenko, B. (1943). Sur la distribution limite du terme maximum d'une série aléatoire. *Ann. Math.*, **44**, 423–453. [Introduced by R.L. Smith and translated by N.L. Johnson in Kotz and Johnson (1992)]

Gumbel, E.J. (1935). Les valeurs extrêmes des distributions statistiques. *Ann. Inst. Henri Poincaré*, **5**, 115–158.

Gumbel, E.J. (1958). *Statistics of Extremes*. Columbia University Press, New York.

Harter, H.L. (1978). *A Chronological Annotated Bibliography of Order Statistics*, **1**. Pre-1950. U.S. Government Printing Office, Washington, DC. [Reprinted by American Sciences Press, Syracuse, NY (1983)]

Johnson, N. L. and Kotz, S. (1997). *Leading Personalities in Statistical Sciences*. Wiley, New York.

Kotz, S. and Johnson, N.L. (1992). *Breakthroughs in Statistics*, Vol. I. Springer, New York.

Mises, R. von (1919). Grundlagen der Wahrscheinlichkeitsrechnung. *Math. Zeitschrift*. [Reprinted in von Mises (1964), Vol. 2]

Mises, R. von (1923). Über die Variationsbreite einer Beobachtungsreihe. *Sitzungsberichte der Berliner Math. Gesellschaft*, **22**, 3–8. [Reprinted in von Mises (1964), Vol. 2]

Mises, R. von (1936). La distribution de la plus grande de *n* valeurs. *Rev. Math. Union Interbalkanique*, **1**, 141–160. [Reprinted in von Mises (1964), Vol. 2]

Mises, R. von (1964). *Selected Papers of Richard von Mises*, Vol. 1 and 2. Amer. Math. Soc., Providence, RI.

Pearson, K. (1902). Note on Francis Galton's difference problem. *Biometrika*, **1**, 390–399.

Tippett, L.H.C. (1925). On the extreme individuals and the range of samples taken from a normal population. *Biometrika*, **17**, 364–387.

Range and Standard Deviation

L. von Bortkiewicz

We use *range* (v) to denote the difference between the largest and the smallest value of a given number (n) of observations of some quantity. Let us focus on the true errors of these observations, i.e., their deviations from the true value, specifically from the mathematical expectation of the quantity in question. If we order these true errors according to their absolute value, i.e., without regard to their sign, beginning with the smallest and concluding with the largest, we obtain a series of positive numbers $\epsilon_1, \epsilon_2 \ldots \epsilon_n$, where $\epsilon_{i+1} \geq \epsilon_i$. It follows that $v = \epsilon_n + \epsilon_{n-1}$ or $v = \epsilon_n + \epsilon_{n-2}$ or $v = \epsilon_n + \epsilon_{n-3}$, etc., up to $v = \epsilon_n + \epsilon_1$ or $v = \epsilon_n - \epsilon_1$, according as already the second largest error has a different sign from the largest or only the third largest or only the fourth largest, etc., or only the smallest, or not a single one of the $n-1$ errors. For an arbitrary symmetrical error law the probability is then $\frac{1}{2}$ that $v = \epsilon_n + \epsilon_{n-1}$, $\frac{1}{4}$ that $v_n = \epsilon_n + \epsilon_{n-2}$, etc., and finally the probability is $1/2^{n-1}$ that $v = \epsilon_n + \epsilon_1$ and with equal probability $1/2^{n-1}$ that $v = \epsilon_n - \epsilon_1$.

If we agree to denote the mathematical expectation of an arbitrary quantity a by $\bar{a}$, we therefore find here that

$$v = \frac{1}{2}(\epsilon_n + \bar{\epsilon}_{n-1}) + \frac{1}{4}(\bar{\epsilon}_n + \bar{\epsilon}_{n-2}) + \cdots + \frac{1}{2^{n-1}}(\bar{\epsilon}_n + \bar{\epsilon}_1) + \frac{1}{2^{n-1}}(\bar{\epsilon}_n - \bar{\epsilon}_1)$$

or

$$\bar{v} = \bar{\epsilon}_n + \frac{1}{2}\bar{\epsilon}_{n-1} + \frac{1}{4}\bar{\epsilon}_{n-2} + \cdots + \frac{1}{2^{n-2}}\bar{\epsilon}_2. \tag{1}$$

From here on the Gaussian error law will be assumed valid. Let the precision be denoted by h and $\frac{2}{\sqrt{\pi}} \int_o^x e^{-t^2} dt$ by $\Theta(x)$ and also by Θ_x. Then $\Theta(hu)$ signifies the probability that an arbitrary error is smaller than u according to its absolute value, and $1 - \Theta(hu)$ the probability that an arbitrary error is larger than u according to its absolute error. Let p_k then denote the probability that an arbitrary error is the kth of n errors and $\theta(u)du$ the probability for an arbitrary error to equal u (i.e., to lie between u and $u + du$). Also let $p_{u,k}$ denote the probability that an error of size u is the kth and finally let $\theta_k(u)du$ denote the probability that the kth error is equal to u.

We have

$$p_k = \frac{1}{n}, \quad \theta(u)du = \frac{2h}{\sqrt{\pi}}e^{-h^2u^2}du,$$

$$p_{n,k} = \binom{n-1}{k-1}\{\Theta(hu)\}^{k-1}\{1-\Theta(hu)\}^{n-k}. \tag{2}$$

Also, because of the relation

$$p_k.\theta_k(u) = \theta(u)du.p_{n,k},$$

we obtain

$$\theta_k(u)du = n\binom{n-1}{k-1}\frac{2h}{\sqrt{\pi}}e^{-h^2u^2}\{\Theta(hu)\}^{k-1}.\{1-\Theta(hu)\}^{n-k}du. \tag{3}$$

Accordingly we find that

$$\bar{\epsilon}_k = n\binom{n-1}{k-1}\int_o^\infty \frac{2h}{\sqrt{\pi}}e^{-h^2u^2}\{\Theta(hu)\}^{k-1}.\{1-\Theta(hu)\}^{n-k}udu,$$

or

$$\bar{\epsilon}_k = \binom{n-1}{k-1}\frac{2n}{h\sqrt{\pi}}\int_o^\infty xe^{-x^2}\Theta_x^{k-1}(1-\Theta_x)^{n-1}dx \tag{4}$$

and, recalling (1), that

$$\bar{v} = \frac{2n}{h\sqrt{\pi}}\int_o^\infty xe^{-x^2}\left\{\Theta_x^{n-1} + (n-1)\Theta_x^{n-2}\left(\frac{1-\Theta_x}{2}\right) + \cdots \right.$$
$$\left. + (n-1)\Theta_x\left(\frac{1-\Theta_x}{2}\right)^{n-2}\right\}dx$$

or

$$\bar{v} = \frac{2n}{h\sqrt{\pi}}\int_o^\infty xe^{-x^2}\left\{\left(\frac{1+\Theta_x}{2}\right)^{n-1} - \left(\frac{1-\Theta_x}{2}\right)^{n-1}\right\}dx. \tag{5}$$

Finally, by means of partial integration, we arrive at the formula

$$\bar{v} = \frac{n(n-1)}{h\pi}\int_o^\infty e^{-2x^2}\left\{\left(\frac{1+\Theta_x}{2}\right)^{n-2} + \left(\frac{1-\Theta_x}{2}\right)^{n-2}\right\}dx. \tag{6}$$

With this expression we have at the same time found the probabilistic relation between the range and the standard deviation (σ). Since $\sigma = 1/h\sqrt{2}$, it follows that

$$\frac{\bar{v}}{\sigma} = \frac{n(n-1)\sqrt{2}}{\pi}\int_o^\infty e^{-2x^2}\left\{\left(\frac{1+\Theta_x}{2}\right)^{n-2} + \left(\frac{1-\Theta_x}{2}\right)^{n-2}\right\}dx. \tag{7}$$

We reach the same result by means of a separation of the positive errors from the negative errors. Let L_m and M_m, respectively, denote the mathematical expectations of the smallest and largest absolute errors among m errors which are now to be regarded as all positive or all negative. By setting $n = m$ and $k = 1$, $k = m$, respectively, we find

$$L_m = \frac{2m}{h\sqrt{\pi}} \int_o^\infty xe^{-x^2}(1 - \Theta_x)^{m-1}dx , \tag{8}$$

$$M_m = \frac{2m}{h\sqrt{\pi}} \int_o^\infty xe^{-x^2}\Theta_x^{m-2}dx. \tag{9}$$

Taking into consideration all possible distributions of the n errors according to their sign, we obtain

$$\bar{v} = \frac{1}{2^n}\left\{2(M_n - L_n) + \sum_{m=1}^{n-1}\binom{n}{m}(M_m + M_{n-m})\right\}$$

or, equivalently,

$$\bar{v} = \frac{1}{2^{n-1}}\left\{-L_n + \sum_{m=1}^{n}\binom{n}{m}M_m\right\}. \tag{10}$$

We only need to substitute in (10) the expressions for L_n and $M_1, M_2, \ldots,$ M_n given by (8) and (9) in order to arrive again at formula (5).

For n not altogether too small, we obtain, as is to be expected, a sufficiently exact value of $\bar{v}$ as follows: instead of considering all possible distributions of the n errors according to their signs, we base the calculation on the most probable distribution, and accordingly express $\bar{v}$, for n even, as $2M_m$, where $m = n/2$. For the determination of M_m, we can also use the formula

$$M_m = \frac{2m(m - 1)}{h\pi} \int_o^\infty e^{-2x^2}\Theta_x^{n-2}dx, \tag{11}$$

which is identical with (9). Then

$$\frac{M_m}{\sigma} = \frac{2m(m - 1)\sqrt{2}}{\pi} \int_o^\infty e^{-2x^2}\Theta_x^{n-2}dx. \tag{12}$$

We obtain from (7) for, e.g., $n = 8$, that $\bar{v}/\sigma = 2.85$ and from (12) that $2M_4/\sigma = 2.92$; for $n = 24$ we find: $\bar{v}/\sigma = 3.90$, $2M_{12}/\sigma = 3.92$.

On the Range of
a Series of Observations

R. von Mises

Allow me to add two remarks to the address by Mr. L. von Bortkiewicz on "Range and Standard Deviation" that was published in these Proceedings.[1] I believe these remarks will usefully supplement the interesting points made by the speaker. One remark includes a simple and general *derivation*, not restricted to the assumption of a Gaussian law, of the formula for the expected value of the range. In the second, I provide an *asymptotic expression* for this expectation, also for a larger class of distributions. For the case of the Gaussian law this expression reduces to a final formula that is very simple, useful in practice, and even instructs us how to find the desired value, without any calculation, in the standard tables of the Gaussian integral.

1

Let $W(x)$ be the probability that the chance-dependent outcome of an observation does not exceed x. Then $W^n(x)$ is the probability that, in n repetitions of the same observation, *none* of the outcomes is larger than x. If now $x_1 < x_2$, then the event whose probability is $W^n(x_2)$ includes two mutually exclusive possibilities: first, that no outcome exceeds x_1 and, second, that the largest observed value lies between x_1 and x_2. Accordingly, $W^n(x_2) - W^n(x_1)$ is the probability that the largest observed value falls in the interval $x_1 < x \leq x_2$.

If we imagine that the entire space of x-values is partitioned into small segments and that the transition to infinitely dense division is carried out, we see that the (Stieltjes) integral

$$g_n = \int x d\left(W^n(x)\right) \tag{1}$$

represents the *expected value of the largest observation*. Here and in the sequel integral signs without specified limits will signify integration from

---

[1] Vol. 21 (1922), pp. 3–11.

$-\infty$ to ∞. Analogously, we find

$$k_n = -\int xd[1 - W(x)]^n \tag{2}$$

as expected value of the *smallest* observed value.[2] It is the difference $g_n - k_n$ that Mr. von Bortkiewicz has termed the expected value of the range v.

If W is differentiable and $w(x)$ the derivative or the "probability density," then it follows from (1) and (2) that

$$\bar{v} = g_n - k_n = n \int w \left[W^{n-1}(x) - (1 - W(x))^{n-1} \right] x dx. \tag{3}$$

In particular, for the Gaussian distribution

$$w(x) = \frac{h}{\sqrt{\pi}} e^{-h^2 x^2}, \ W(x) = \frac{h}{\sqrt{\pi}} \int_{-\infty}^{x} e^{-h^2 x^2} dx = \frac{1}{2}[1 + \Theta(hx)] \tag{4}$$

we have, setting $hx = z$ and writing Θ_z for $\Theta(z)$,

$$\bar{v} = \frac{n}{h\sqrt{\pi}} \int ze^{-z^2} \left[\left(\frac{1 + \Theta_z}{2} \right)^{n-1} - \left(\frac{1 - \Theta_z}{2} \right)^{n-1} \right] dz. \tag{5}$$

This is the result that is expressed in Mr. von Bortkiewicz' formulae (5)–(7).

2

In order to obtain an asymptotic representation of v it suffices to seek this for g_n, as defined in (1). We prove the following theorem. *Let $W(x)$ be a monotone continuous function, increasing from 0 to 1 and behaving at $\pm\infty$ in such a way that $\int |x| dW(x)$ exists and is finite.*[3] *Further, suppose that, for an arbitrary positive fixed ξ, the ratio $[1 - W(x + \xi)]/[1 - W(x)]$ tends to 0 as x increases, and let x_n denote the (unique) value of x for which*

$$W(x_n) = 1 - \frac{1}{n}. \tag{6}$$

Then

$$\lim_{n\to\infty} \frac{g_n}{x_n} = \lim_{n\to\infty} \frac{1}{x_n} \int xd \left(W^n(x) \right) = 1. \tag{7}$$

---

[2] However, it is to be noted that, for a piecewise continuous W, the limits in (2) must be taken from the left in (2) and from the right in (1).

[3] This requirement is equivalent to the condition that the integrals $\int_{-\infty}^{o} W dx$ and $\int_{o}^{\infty} (1 - W) dx$ be finite; for the second of these integrals the condition is already contained in the second assumption of the theorem.

In order to prove this, we first recast the integral (1) in such a way that the region of integration is divided into two segments by an intermediate point x_n. Applying partial integration to each segment, we obtain (and do so for an arbitrary x_n)

$$\int x\,d(W^n) = x_n - \int_{-\infty}^{x_n} W^n dx + \int_{x_n}^{\infty} (1 - W^n)dx. \tag{8}$$

In fact, both sides of (8) represent the area of the region enclosed by the y-axis, the curve $y = W^n(x)$, and the asymptotes $W = 0$ and $W = 1$, respectively. We may confine ourselves to the case where W assumes the value 1 only at infinity; for if this were to occur already for $x = X$, then, as is easily seen, X would be the limit for x_n as well as for g_n, so that our claim (7) is certainly true. In other words: if X is the largest *possible* observed value, then this is also the expected value of the largest of n observations for sufficiently large n.

The quantities x_n defined by (6) go to infinity with increasing n, so that we have at once

$$\lim_{n \to \infty} W^n(x_n) = \lim \left(1 - \frac{1}{n}\right)^n = e^{-1}. \tag{9}$$

We now introduce the ratio

$$\frac{1 - W(x_n + \xi)}{1 - W(x_n)} = n[1 - W(x_n + \xi)] = \rho \tag{10}$$

which, by our assumption on the behavior of W for increasing n, goes to 0 or ∞ according as ξ is positive or negative. Then

$$W^n(x_n + \xi) = \left(1 - \frac{\rho}{n}\right)^n \le e^{-\rho} \to 0 \quad \text{for} \quad \xi < 0, \tag{11a}$$

$$1 \quad W^n(x_n + \xi) = 1 - \left(1 - \frac{\rho}{n}\right)^n \le 1 - (1 - \rho) \to 0 \quad \text{for} \quad \zeta > 0. \tag{11b}$$

Graphically stated, this means that the curve $W^n(x)$, the curve of the distribution of the largest observed value, approaches with increasing n the *rectangular path* consisting of the x-axis from $-\infty$ to the point x_n, the vertical at x_n to height 1, and finally the horizontal at height 1 up to $x = \infty$. We now wish to see that the two integrals on the right of (8), which represent the two areas between the W^n-curve and the straight-line path just described, go to zero with increasing n, apart from a vanishingly small multiple of x_n.

We split up the first integral into four parts by means of three intermediate points with abscissae $-X$, 0, $x_n - \xi_1$. First we choose $X > 0$ in such a way that the integral of W from $-\infty$ to $-X$ is smaller than an arbitrary preassigned ϵ. This is always possible in view of the first assumption regarding the behavior of W. Then the integral of W^n over the same region is certainly still smaller than ϵ, since everywhere here $W < 1$. For the same reason the integral of W^n extending from $-X$ to 0, which by the first mean value theorem is at most equal to $XW^n(0)$, can be made arbitrarily small. The integral of W^n from 0 to $x_n - \xi_1$, where ξ_1 is any positive quantity, is by this same theorem at most equal to $x_nW(x_n - \xi_1)$; by (11a) it is therefore less than any arbitrarily small multiple of x_n. Finally, we choose $\xi_1 < \epsilon$, in order that, because of $W \leqq 1$, the fourth piece of the integral also becomes $\leqq \epsilon$.

By introducing the new variable of integration $\xi = x - x_n$, we write the second integral on the right side of (8) as

$$\int_o^\infty [1 - W^n(x_n + \xi)]\, d\xi. \tag{12}$$

In every finite region, e.g., from 0 to 1, we can, on account of (9) and (11b), make the integral arbitrarily small by increasing n. In view of the infinite upper terminal we observe, by the second assumption on W, that $1 - W(x)$ must tend to zero faster than exponentially. Or stated more precisely: let ρ_1 denote the value of the ratio (10) for $\xi = 1$ (which therefore depends only on n and vanishes with increasing n). Then, by successively following the points $x_n + 1, x_n + 2, \ldots$ we obtain, for sufficiently large x_n,

$$1 - W(x_n + \xi) \leqq [1 - W(x_n)]\rho_1^{\xi-1} = \frac{1}{n}\rho_1^{\xi-1}.$$

But from this it follows that for $\rho_1 < 1$, $\xi \geqq 1$,

$$W^n(x_n + \xi) \geqq \left(1 - \frac{1}{n}\rho_1^{\xi-1}\right)^n \geqq 1 - \rho_1^{\xi-1}$$

and that the still remaining part of the integral (12) is smaller than

$$\int_1^\infty \rho_1^{\xi-1}\, d\xi = -\frac{1}{\ln \rho_1},$$

and so vanishes together with ρ_1.

Thus we have proved eq. (7) and our theorem. Moreover, not only the expected value, i.e., the average of many observations of g_n, lies near x_n, but also *with a probability approaching 1 it may be assumed that, for large n, the largest of n observations falls near x_n*. We can also give the number x_n a clear probabilistic interpretation. According to Bernoulli's Theorem it may be expected "almost surely," for fixed m and sufficiently large n, that $n(1 - 1/m)$ observations lie to the left of x_m. If this statement is extended to $m = n$, which is not automatically permissible, it means that the largest observation will lie almost surely to the right of x_n. Our theorem means, in addition, that the largest observation may be expected to be *very near and almost surely* close to x_n.

3

The Gaussian distribution (4) certainly fulfills both the conditions of our theorem, the first being here included in the second by the symmetry. That the latter is satisfied is seen from

$$\lim_{x \to \infty} \frac{1 - W(x + \xi)}{1 - W(x)} = \lim_{x \to \infty} \frac{w(x + \xi)}{w(x)}$$

$$= \lim_{x \to \infty} e^{-2x\xi - \xi^2} = \begin{cases} 0 & \text{for} \quad \xi > 0 \\ \infty & \text{for} \quad \xi < 0 \end{cases}. \quad (13)$$

In addition, we have in view of the symmetry that $g_n = -k_n$, $\bar{v} = 2g_n$. Also, according to (13), for negative ξ the ratio ρ defined in (10) tends to infinity as fast as e^x, so that $xe^{-\rho}$ tends rapidly to zero. Hence we see that the dominant term in the integral of (8), shown above only to be small relative to x_n, vanishes here even compared to 1. Thus, for the Gaussian distribution, we have

$$\lim (g_n - x_n) = 0, \quad (14)$$

a stronger result than (7); in fact, the difference tends to zero more rapidly than any power of $1/n$.

The value of x_n, which by (4) and (6) is now given by $\Theta(hx_n) = 1 - 2/n$, can be taken from a table of Θ, with $hx_n = z$. Since $h = 1/\sigma\sqrt{2}$, we obtain the following result for the ratio of the expected value of range to standard deviation

$$\frac{\bar{v}}{\sigma} = 2\sqrt{2}z; \quad \Theta(z) = 1 - \frac{2}{n}. \quad (15)$$

For example, we find from a 7-digit table of the Gaussian integral

$$\frac{\bar{v}}{\sigma} = \quad 6.18 \quad 7.44 \quad 8.53 \quad 9.50 \quad 10.41$$

for

$$n = \quad 10^3 \quad 10^4 \quad 10^5 \quad 10^6 \quad 10^7.$$

In comparison, the exact calculation of Mr. von Bortkiewicz yields in the first three cases: 6.48, 7.70, and 8.70.

If we wish to express $\bar{v}/\sigma$ as a function of n, we need only use the well-known asymptotic expansion of Θ and solve for z in the equation

$$\frac{1}{n} = \frac{e^{-z^2}}{2z\sqrt{\pi}} \left[1 - \frac{1}{2z^2} \cdots \right] \quad \text{or} \quad z^2 - \ln \frac{n}{2\sqrt{\pi}} = -\ln z - \frac{1}{2z^2} \cdots . \quad (16)$$

This yields as first approximation

$$z_0 = \sqrt{\ln \frac{n}{2\sqrt{\pi}}}, \quad \text{hence} \quad \frac{\bar{v}}{\sigma} = 2\sqrt{2 \ln \frac{n}{2\sqrt{\pi}}} \quad (17)$$

and as a second, in practice entirely adequate approximation

$$z = z_0 \left(1 - \frac{\ln z_0}{2 z_0 + 1} \right), \quad \text{hence} \quad \frac{\bar{v}}{\sigma} = 2\sqrt{2} z_0 \left(1 - \frac{\ln z_0}{2 z_0 + 1} \right). \quad (18)$$

P. Frank[4] has used expression (17), without entering into the proof of eq. (14), in order to calculate the mean velocity of the "head" of a diffusing particle swarm. From a probabilistic point of view this velocity is identical with the quantity investigated here.

---

[4] *Phys. Zeitschr.* 19 (1918), 516–520.

The Evaluation of Tournament Outcomes

Comments on Zermelo (1929)

1 Introduction

This paper by the noted German mathematician Ernst Zermelo (1871–1953) was long overlooked and was brought to the attention of the statistical community only in the mid-1960s, by John Moon and Leo Moser, professors of mathematics at the University of Alberta, Canada. Zermelo is concerned with the evaluation of players in chess tournaments, especially for tournaments lacking the balance of Round Robins, where all pairs of players meet equally often. There had long been an obvious method for dealing with Round Robins, namely to rank players according to their number of wins (counting draws as half-wins). But what was to be done in the absence of balance? Zermelo proposes that a "playing strength" $u_r (\geq 0)$ be associated with player A_r, $r = 1, \ldots, n$, such that the probability of A_r defeating A_s is given by

$$u_{rs} = \frac{u_r}{u_r + u_s}. \tag{a}$$

The probability model (a) was independently introduced by Bradley and Terry (1952) and has become well known as the Bradley–Terry paired-comparison model. Although the models in the two papers are identical, it is fascinating to observe the differences in emphasis. Zermelo is solely concerned with a very thorough examination of his method of estimation — actually maximum likelihood — for the most general tournament. He never gives a thought to applications beyond tournaments.

The statisticians Bradley and Terry do not even mention tournaments, but are motivated by paired-comparison experiments in which, as in taste-testing, n "treatments" are compared in pairs by one or more "judges." Confining themselves to the balanced case, they immediately apply their techniques to experimental data and provide some helpful tables for the analysis. New problems arise, due to possible differences between the judges, and in addition to estimation the authors also consider several tests of hypotheses.

This is not the end of the story of rediscovery. Apparently unaware of both the above papers, Ford (1957), another mathematician, is concerned with "ranking a collection of objects on the basis of a number of binary comparisons." Like Zermelo, he is motivated by the unbalanced case and concisely recaptures many of Zermelo's results. With Bradley and Terry he shares an awareness of the psychometric literature on "the method of paired comparisons," a term coined by Thurstone (1927), who pioneered the theory of the subject.

As we will see, Zermelo's paper, apart from its historic interest, contains some results that even now have apparently not been incorporated into the paired-comparison literature. This applies especially to his iterative method of solving the likelihood equations for the u_r of (a).

2   Main Results

In this section we outline and discuss Zermelo's main results, leaving some interesting special issues to the following sections. This includes the treatment of drawn games that are here assumed absent.

Zermelo arrives swiftly at his model (1) by requiring the playing strength u_r of $A_r(r = 1, \ldots, n)$ to be proportional to A_r's chance of winning an individual game. This proportionality assumption is simple and natural. Although the resulting model has proved widely useful, note that it can not cope with probabilistic intransitivities such as $P(A_1 \to A_2) > \frac{1}{2}$, $P(A_2 \to A_3) > \frac{1}{2}$, $P(A_3 \to A_1) > \frac{1}{2}$, where $\to$ means "defeats." (Of course, the occasional observed intransitivity, or "circular triad," $A_1 \to A_2 \to A_3 \to A_1$ causes no problems.)

Zermelo immediately considers the most general tournament and writes down the probability of the observed outcome in (3), apart from a constant factor that he consciously omits. His Π' should extend over $r < s$, not $r \neq s$, but he *uses* Π' correctly later. To estimate the parameters he maximizes Φ in (3), unaware of the method of maximum likelihood, by then well established. Instead, he calls on a method "often used in the theory of gases"!

There follows a careful discussion of the maximization of Φ (but in (6) it would have been more logical to have written $\Phi(\frac{1}{n}, \ldots, \frac{1}{n})$ instead of $\Phi(1, \ldots, 1)$). It is shown that a general tournament can be uniquely decomposed into subtournaments that are irreducible in the sense that they can not be divided into two parts, such that no player in the first defeats any player in the second.

In the irreducible case all u_r are shown to be strictly positive and uniquely determined up to a proportionality constant. This is done in his Section 2. For reasons of space we have omitted Section 3 which deals with the general case. However, the not unexpected results can be easily described in slightly different terms than used by Zermelo. For each irreducible subtournament C_i ($i = 1, \ldots, m$, say) determine the corresponding u_r by maximum likelihood as in Section 2. For A_r and A_s both in C_i, $P(A_r \to A_s) = u_{rs}$ is then positive and uniquely determined. If $A_r \in C_i$, $A_s \in C_j$, where $C_i < C_j$, i.e., no player in C_i defeats any player in C_j, then $u_{rs} = 0$. Finally A_r and A_s may not be comparable because neither has lost to the other, even indirectly, i.e., through a chain of intermediaries. Zermelo gives a simple example for three players A_1, A_2, A_3: if A_1 and A_2 have both defeated or both lost to A_3, then, if they have not met, they are not comparable. This leaves u_{12} indeterminate.

In his Section 4, Zermelo focuses on "regular tournaments," i.e., Round Robins. By a simple argument he establishes the now well-known result that his method ranks the players in the same order as does the number of wins. But later writers (including this one) seem to have missed his stronger result (29) showing that, with $\Sigma u_r = 1$, the u_r are more widely spaced than the equally normed $g'_r = g_r / \frac{1}{2}kn(n-1)$, provided $n > 2$. A case for the desirability of wider spacing is made in the introduction of the paper. But it is also clear that the difference between two players tends to be underestimated by the difference in their g'_r values through the presence of much stronger and/or much weaker players against whom they both lose or both win.

The numerical solution of the likelihood equations for the u_r by an iterative method is treated in detail in Zermelo's Section 5. Bradley and Terry (1952) give a similar method in the balanced case, but one involving constant updating of the estimates. They are content to find that the method works in their experience. Ford (1957) uses exactly the same method as Zermelo and discusses convergence. However, Zermelo proves a more general result and also shows that the convergence is at least as fast as that of a geometric series (eq. (54)).

3   Landau and the Kendall–Wei Method

Zermelo's first footnote reference to a 1914 paper by Edmund Landau is very interesting. It shows that another eminent mathematician had considered problems arising in the evaluation of chess tournaments and had pointed out already the difficulties with a method apparently advanced independently by Wei (1952) and popularized by Kendall (1955).

This approach, confined to Round Robin tournaments, is an attempt to improve on the usual scoring system of 1 point for a won game, $\frac{1}{2}$ point for a draw, and 0 for a loss. The basic idea is to reward a player more for defeating a strong opponent than a weak one. For an unreplicated Round Robin the tournament matrix $\mathbf{A} = (a_{rs})$ has $a_{rs} + a_{sr} = 1$ for $r \neq s$, with $a_{rs} = 1$, $\frac{1}{2}$, or 0, and $a_{rr} = 0$ $(r, s = 1, \ldots, n)$. The usual row-sum method gives A_r's score as $g_r = \Sigma_s g_{rs} = (\mathbf{A1})_r$, the sum of the rth row of $\mathbf{A}$. Greater rewards for defeating strong opponents are reflected in scores $g_r^{(i)} = (\mathbf{A}^i \mathbf{1})_r$, $i = 1, 2, \ldots$

Already in 1895 Landau pointed out inconsistencies in this matrix-powering approach and suggested that the score-vector **s** should satisfy (in our notation)

$$\mathbf{A}\mathbf{s} = \lambda\mathbf{s}, \tag{b}$$

where $\lambda \geq 0$. By 1914 he was able to take advantage of the recently developed Perron–Frobenius theory for nonnegative matrices to discuss eq. (b) for irreducible matrices $\mathbf{A}$.

Of course, (2) applies also to replicated Round Robins if a_{rs} is interpreted as the proportion of A_r's wins against A_s. Landau shows, however, for the following two tournament outcomes

		A_1	A_2	A_3
	A_1		1	$1-\epsilon$
(I)	A_2	0		$\frac{1}{2}$
	A_3	ϵ	$\frac{1}{2}$	

		A_1	A_2	A_3
	A_1		1	$1-\epsilon$
(II)	A_2	0		1
	A_3	ϵ	0	

where $0 < \epsilon < 1$, that for suitable choice of ϵ, player A_2 receives a lower score in (II) than in (I). This leads Landau to dismiss the Kendall–Wei approach 40 years prior to its birth!

Still, Kendall–Wei or matrix-powering methods are useful, at least, in often breaking ties between players with the same number of wins. In addition to Kendall (1955), see also Moon (1968, p. 42) and David (1988, p. 104).

4   The Treatment of Drawn Games (Ties)

Zermelo naturally treats a drawn game as half a win for each of the two players. This is unexceptional as a scoring system, but he does not seem to realize that his model no longer applies. Since the probability that A_r and A_s will draw their game is according to (a)

$$\frac{\sqrt{u_r u_s}}{u_r + u_s},$$

we see that the probability of A_r winning, drawing, or losing exceeds 1.

The situation is in fact the special case $\nu = 1$ of Davidson's (1970) extension of the Bradley–Terry model, which in present notation takes

$$u_{rs} = \frac{u_r}{u_r + u_s + \nu\sqrt{u_r u_s}}, \quad u_{sr} = \frac{u_s}{u_r + u_s + \nu\sqrt{u_r u_s}}$$

and the probability of a draw as

$$\frac{\nu\sqrt{u_r u_s}}{u_r + u_s + \sqrt{u_r u_s}}.$$

In this model the parameter ν reflects the ease with which a draw can be achieved between two players of given strengths.

Davidson shows how ν and the u_r may be estimated from the data and also that for Round Robins the estimates of the u_r are in the same order as the usual scores. See also David (1988, p. 137).

5   A Note on the Author

Born in Berlin in 1871, Ernst Zermelo was largely educated there, but attended also Halle and Freiburg Universities. Upon completion, in 1894, of his Berlin doctoral thesis on the calculus of variations, he became an assistant to Max Planck, then professor of theoretical physics at Berlin. Under Planck, Zermelo worked in hydrodynamics. However, after a move to Göttingen in 1897, he changed course and began his deep study of set theory, making the fundamental contributions for which he became well known.

Poor health forced Zermelo to leave a chair in mathematics at Zürich after just six years, in 1916. The present paper was written in Freiburg, Germany, where he had been appointed to an honorary chair in 1926. Earlier, in 1913, he had applied set theory to the theory of chess. Zermelo renounced his chair in 1935, in disapproval of Hitler's regime. At his request, he was reinstated in 1946! He died in Freiburg in 1953.

This note is based on a much fuller account by Rootselaar (1981).

References

Bradley, R.A. and Terry, M.E. (1952). The rank analysis of incomplete block designs. I. The method of paired comparisons. *Biometrika*, **39**, 324–345.

David, H.A. (1988). *The Method of Paired Comparisons*, 2nd edn. Griffin, London.

Davidson, R.R. (1970). On extending the Bradley–Terry model to accommodate ties in paired comparison experiments. *J. Amer. Statist. Assn.*, **65**, 317–328.

Ford, L.R., Jr. (1957). Solution of a ranking problem from binary comparisons. *Amer. Math. Monthly*, **64**, 28–33.

Kendall, M.G. (1955). Further contributions to the theory of paired comparisons. *Biometrics*, **11**, 43–62.

Moon, J.W. (1968). *Topics on Tournaments*. Holt, Rinehart, and Winston, New York.

Rootselaar, B. van (1981). Zermelo. *Dictionary of Scientific Biography*, **13**, 613–616.

Thurstone, L.L. (1927). A law of comparative judgment. *Psychol. Rev.*, **34**, 273–286.

Wei, T.H. (1952). The algebraic foundations of ranking theory. Unpublished Thesis, Cambridge University.

Zermelo, E. (1929). Die Berechnung der Turnier-Ergebnisse als ein Maximumproblem der Wahrscheinlichkeitsrechnung. *Math. Zeit.*, **29**, 436–460.

The Evaluation of Tournament Results as a Maximization Problem in Probability Theory

E. Zermelo

Chess tournaments are of late always arranged so that each player meets every other player k times, where k is fixed (usually $k = 2$, with alternate colors). The ranking of the players is then determined by the number of games won, where an undecided ("drawn") game is counted as half a win for each of the partners. This method of calculation treats all games equally, regardless of the order of play, and consequently reduces chance effects to a minimum. Apparently the procedure has worked very well in practice if only a *ranking* is of interest, but it fails completely for *broken-off* tournaments (in which the number of games played is not the same for each participant).

Also, the method does not provide a satisfactory *quantitative* determination of the *relative strengths* of the players, since it evidently favors mediocrity. For example, for $k = 1$, a chess master would win only n games against n duffers, but together these would win $n(n-1)/2$ games, each $(n-1)/2$ on average. The ratio of these values, $n : (n-1)/2 = 2n/(n-1)$ would approach 2 as n increases, so that the winner would be rated only twice as high as each of the other players.

Various proposals have been made and discussed in the chess literature for removal of this weakness and for quantitative improvement of the usual procedure, e.g., by F. Landau.[1] However, interesting as these proposals were from a mathematical point of view, they have led occasionally to paradoxical results and so far have provided no satisfactory solution of the problem. In the sequel, a new procedure for the evaluation of tournaments will be developed which appears to be free of previous objections. Also, especially for "broken-off" and "combined" tournaments, the method leads to mathematically determined results that satisfy all sensible requirements.

Our procedure amounts to regarding the relative strengths of the players as probabilities that are to be determined so that the probability of the occurrence of the observed tournament results is maximized.

---

[1]E. Landau. On the distribution of prizes in tournaments. *Ztschr. f. Math. u. Phys.* **63**, p. 192 (1914).

1   Statement and Reduction of the Problem

We regard a positive number u_r attached as "playing strength" to each of n players A_r participating in a tournament and we suppose that these numbers $u_1, u_2, \ldots, u_n$ are in the same relation as their relative chances of winning. The probability that in a particular game A_r wins against A_s is therefore given by the fraction

$$u_{rs} = \frac{u_r}{u_r + u_s} \, . \tag{1}$$

Let k_{rs} be the number of games played by A_r against A_s, where we suppose here that no game ends in a draw,[2] with g_{rs} won by A_r and $g_{sr} = k_{rs} - g_{rs}$ won by A_s. If we regard the various games as "independent" events, then the combined probability of the partial result that A_r wins exactly g_{rs} games against A_s, and A_s accordingly g_{sr} against A_r, is

$$w_{rs} = u_{rs}^{g_{rs}} u_{sr}^{g_{sr}} = \frac{u_r^{g_{rs}} u_s^{g_{sr}}}{(u_r + u_s)^{k_{rs}}} \, . \tag{2}$$

Omitted in (2) is a numerical factor, independent of u_r and u_s, namely $\sigma_{rs} = \binom{k_{rs}}{g_{rs}}$, which Professor Ostrowski has kindly drawn to my attention. Consequently, the probability of the overall result, determined by the matrix g_{rs}, is the product of all these probabilities:

$$w = \prod_{r,s}' w_{rs} = \frac{u_1^{g_1} u_s^{g_2} \ldots u_n^{g_n}}{\prod_{r,s}' (u_r + u_s)^{k_{rs}}} = \Phi(u_1, \ldots, u_n) \, , \tag{3}$$

where

$$g_r = \sum_{s}^{s \neq r} g_{rs} \tag{4}$$

always denotes the total number of games won by A_r and where the product $\prod'$ extends over all combinations r, s, with $r \neq s$.

The (nonnegative) numbers $u_1, \ldots, u_n$ are now to be determined in such a way that $\Phi(u_1, \ldots, u_n)$ becomes as large as possible, so that the observed tournament result, determined by the numbers $g_1, \ldots, g_n$, becomes most probable.[3]

---

[2] For all practical applications we take any undecided ("drawn") games into account by means of the fiction, that twice as many games were played; for each player we count each victory as two wins and each draw as one win. This corresponds exactly to the procedure in the introduction of regarding drawn games as half-wins.

[3] This approach is analogous to a procedure frequently used, especially in the theory of gases.

Since Φ is homogeneous, of dimension zero, in the variables $u_1, \ldots, u_n$, and therefore dependent only on their ratios, we may assume without loss of generality that

$$\sum_{r=1}^{n} u_r \leq 1 , \tag{5a}$$

while at the same time, for each r,

$$u_r \geq 0 \qquad (r = 1, 2, \ldots, n). \tag{5b}$$

The region $\mathcal{B}$ of the u_r, determined by (5a) and (5b), is finite and closed. Inside $\mathcal{B}$ the function Φ, defined as a probability by (3), is everywhere ≤ 1, and therefore has a finite upper bound $\bar{w} \leq 1$. It follows that

$$\bar{w} \geq \Phi(1, \ldots, 1) = \left(\frac{1}{2}\right)^{N} > 0 , \tag{6}$$

where

$$N = \sum_{r=1}^{n} g_r = \sum_{r,s}^{r \neq s} k_{rs} \tag{7}$$

is the total number of games actually played. According to Weierstrass, $\mathcal{B}$ contains at least one point $\bar{q}$, with coordinates $\bar{u}_1, \ldots, \bar{u}_n$, in whose neighborhood the value $\bar{w}$ is approximated by Φ.

We must now distinguish the two cases of $\bar{q}$ lying in the interior or on the boundary of $\mathcal{B}$.

Case 1. Suppose $\bar{q}$ is in the interior of $\mathcal{B}$; i.e.,

$$\sum_{r=1}^{n} \bar{u}_r < 1 \text{ and } \bar{u}_r > 0 \qquad (r = 1, 2, \ldots, n) .$$

Then Φ is *continuous* at the point $\bar{q}$ and

$$\Phi(\bar{q}) = \bar{w}$$

is the *maximum* of Φ in $\mathcal{B}$. Since Φ is then differentiable with respect to all variables and is nonzero, the partial derivatives of $\log \Phi$ must all vanish at $\bar{q}$; i.e.,

$$\frac{g_r}{\bar{u}_r} - {\sum_t}' \frac{k_{rt}}{\bar{u}_r + \bar{u}_t} = 0$$

or

$${\sum_t}' k_{rt} \frac{\bar{u}_r}{\bar{u}_r + \bar{u}_s} = g_r \quad \text{for} \quad r = 1, \ldots, n , \tag{8}$$

where the sum $\sum'$ is to be taken over all $t \neq r$. We obtain therefore a set of n homogeneous algebraic equations for the determination of $\bar{u}_1, \ldots, \bar{u}_n$ or rather for their ratios $\bar{u}_r : \bar{u}_s$.

By addition of the n equations (8), we obtain, because of

$$\frac{\bar{u}_r}{\bar{u}_r + \bar{u}_t} + \frac{\bar{u}_t}{\bar{u}_r + \bar{u}_t} = 1 ,$$

the identity (7) as an expression of the dependence between the equations of the system.

Since $k_{rt}\frac{u_r}{u_r+u_t}$ represents the "probable number" of games won by A_r against A_t, the system of equations (8) means that, for the evaluation of the players according to our approach, *the probable number of games to be won by one player against all the other players is equal to the number actually won by him.*

Case 2. Suppose $\bar{q}$ lies on the *boundary* of $\mathcal{B}$ in such a way that all

$$\bar{u}_r > 0 , \qquad (r = 1, 2, \ldots, n)$$

but

$$\sum_r \bar{u}_r = 1 .$$

Since the function Φ is homogeneous in $u_1, \ldots, u_n$ and of dimension zero, we have

$$\Phi(\lambda u_1, \ldots, \lambda u_n) = \Phi(u_1, \ldots, u_n) . \tag{9}$$

Hence, for $0 < \lambda < 1$, the boundary value $\bar{w}$ is also approximated in the *interior* of $\mathcal{B}$ which reduces Case 2 to Case 1.

Case 3. Suppose the point $\bar{q}$ lies on the boundary of $\mathcal{B}$ in such a way that

$$\prod \bar{u}_r = \bar{u}_1 \bar{u}_2 \ldots \bar{u}_n = 0; \tag{10}$$

i.e., at least one $\bar{u}_r = 0$.

If, e.g., $\bar{u}_1 = 0$, with all other $\bar{u}_r > 0$, then the denominator of the function Φ is different from zero at $\bar{q}$. Therefore we must also have $g_1 = 0$, since according to (3) Φ would contain as a factor a power $u_1^{g_1}$ of u_1. Consequently, Φ instead of equaling $\bar{w} > 0$ would approximate zero. Conversely, if $g_1 = 0$, then u_1 appears in (3) only in the denominator, namely in all the positive factors $u_1 + u_s$. So in the ascent to the upper limit $\bar{w}$, u_1 must tend to its lower limit 0; i.e., $\bar{u}_1 = 0$. According to our approach a single player will be rated with playing strength zero if and only if he is the only one in the tournament to lose all his games.[4]

In general,[5] for suitably chosen labeling, let

$$\bar{u}_1 = \bar{u}_2 = \cdots = \bar{u}_p = 0; \quad \bar{u}_{p+1}, \bar{u}_{p+2}, \ldots, \bar{u}_n > 0 .$$

---

[4]This result is only apparently paradoxical, since no player could do worse, however low his playing strength. Cf. Landau, loc. cit., p. 201.

[5]The case $\bar{u}_r = 0$ for all r can be excluded because of the homogeneity (9) of Φ.

Then it follows, as in the special case, that Φ in (3) can not contain a factor of the form $u_{rs} = \frac{u_r}{u_r + u_s}$ for $r \leq p < s$; for otherwise this factor, and hence Φ, would have to tend to zero as the boundary point $\bar{q}$ is approached, which is impossible in view of (6): $\bar{w} > 0$. Thus, in this case, $g_{rs} = 0$ for all $r \leq p < s$; i.e., we have a *partition of all players into two classes, such that no player A_r in the first class wins a game against any player in the second class* (but loses all such games).

Conversely, suppose the result of the tournament has led to such a class formation $\{A_r, A_s\}$, with $g_{rs} = 0$. *Then as Φ approaches $\bar{w}$, all u_r assigned to the players A_r of the inferior class, which for at least one $s > p$ satisfy the condition $k_{rs} > 0$, must tend to zero:* $\bar{u}_r = 0$.

In order to prove this, we decompose the function defined by (3) into three factors

$$\Phi(u) = \Phi_1(u_r)\Phi_2(u_s)\Phi_{1,2}(u_r, u_s) , \tag{11}$$

where

$$\Phi_1(u_r) = \prod_{1 \leq r,r' \leq p}^{r \neq r'} \frac{u_r^{g_{rr'}} u_{r'}^{g_{r'r}}}{(u_r + u_{r'})^{k_{rr'}}} ,$$

$$\Phi_2(u_s) = \prod_{p+1 \leq s,s' \leq n}^{s \neq s'} \frac{u_s^{g_{ss'}} u_{s'}^{g_{s's}}}{(u_s + u_{s'})^{k_{ss'}}} ,$$

$$\Phi_{1,2}(u_r, u_s) = \prod_{r=1}^{p} \prod_{s=p+1}^{n} \left(\frac{u_s}{u_r + u_s}\right)^{k_{rs}} ,$$

since, by assumption, factors u_{rs} do not appear at all because of $g_{rs} = 0$.

If we now replace all u_r, with $r \leq p$, by λu_r, $0 < \lambda < 1$, then Φ_1 and Φ_2 remain unchanged, being homogeneous, whereas $\Phi_{1,2}$ becomes

$$\Phi_{1,2}^{(\lambda)} = \prod_{r=1}^{p} \prod_{s=p+1}^{n} \left(\frac{u_s}{\lambda u_r + u_s}\right)^{k_{rs}} = \Phi_{1,2}(\lambda u_r; u_s) . \tag{12}$$

Suppose that at a point $\bar{q}$, where $\Phi(\bar{q})$ approximates $\bar{w}$, one of the values $\bar{u}_r$ is positive $(r \leq p)$, with $k_{rs} > 0$. Then, since $\lambda > 0$,

$$\frac{\Phi_{1,2}^{(\lambda)}}{\Phi_{1,2}} \geq \left(\frac{u_r + u_s}{\lambda u_r + u_s}\right)^{k_{rs}} ,$$

where the ratio on the right, in view of $\bar{u}_r > 0$, $\lambda < 1$, and $k_{rs} > 0$ tends to the value

$$\left(\frac{\bar{u}_r + \bar{u}_s}{\lambda \bar{u}_r + \bar{u}_s}\right)^{k_{rs}} > 1$$

as $q \to \bar{q}$. Therefore at $\bar{q}$ we would have

$$\Phi_{1,2}^{(\lambda)} > \Phi_{1,2}$$

and accordingly

$$\lim \Phi(\lambda u_1, \ldots, \lambda u_p; u_{p+1}, \ldots, u_n) > \lim \Phi(u_1, \ldots, u_n) = \bar{w} , \tag{13}$$

in contradiction to the definition of $\bar{w}$.

It follows therefore that for $r \leqq p$ all those $\bar{u}_r$ must vanish for which at least one $k_{rs} > 0$.

As q approaches $\bar{q}$, $\Phi_{1,2} \to 1$, since here only those u_r with $k_{rs} > 0$ actually appear for which $\bar{u}_r = 0$. Accordingly, $\bar{w}$ is approximated by Φ if and only if $\Phi_1(u_r)$ and $\Phi_2(u_s)$ become *as large as possible, independently of each other.*

In order to determine the relative strength of players in the same class, it suffices therefore to apply our approach to the subtournament consisting of the players in this class. Our problem is thereby reduced to tournaments with smaller numbers of players. These are again subject to similar reductions, until "irreducible" problems are reached for which all $\bar{u}_r$ are positive, so that the solution given by equation (8) is valid.

We will show first that the decomposition of the "total tournament" T into "irreducible" subtournaments or "prime tournaments" $C_1, C_2, C_3, \ldots$ is unique. The group C of players in such a "prime tournament" is in fact characterized by the property that all other players in T fall into two groups P and Q,

$$T = P + C + Q , \tag{14}$$

such that *no* player in P wins a game against one in C or Q and *no* player in C wins against one in Q. On the other hand, *within* C no division into two classes C', C'' with the stated property is possible.

Now if

$$T = T' + T''$$

is any division of the total tournament, such that no player in T' wins against one in T'', then at the same time P and Q each split into two classes

$$P = P' + P'', \quad Q = Q' + Q'' ,$$

where P' and Q' belong to T' and P'' and Q'' to T'', whereas C belongs *entirely* to T' or entirely to T'', since otherwise

$$C = C' + C''$$

would be divisible, contrary to assumption. If therefore, in continued decomposition of the tournament, the part that has even one element in common with C is further divided, then the ultimately remaining indivisible portion C^* must contain all of C and conversely C must contain C^*; i.e., $C = C^*$ and the decomposition (into prime tournaments) is consequently *unique.*

If C_1 is any prime tournament different from C, it must lie *entirely* in P or *entirely* in Q, since otherwise it would be divisible. If no games at all have been played between C and C_1, then C_1 lies *arbitrarily* in either P or Q.

2   The Unique Solution of the Irreducible Problem

We first restrict ourselves to the "irreducible" case. Thus we assume that a division of all the players into two classes is *impossible*, where no player A_r of the inferior class wins even one game against one A_s of the superior class, so that $g_{rs} = 0$ for every such combination r, s. Then, as follows from the arguments in Section 1, for *no* approximation to the upper bound $\bar{w}$ does any of the ratios u_r/u_s approach zero (when some, but not all, $\bar{u}_r$ would be zero). In fact, only Case 1 or Case 2 can arise, where $\bar{w}$ is actually the *true maximum* of the function Φ for a system of *positive* values $u_t = \bar{u}_t > 0$. For such values u_t equations (8) therefore also hold and we write them now in the form

$$\sideset{}{'}\sum_t k_{rt} \frac{u_r}{u_r + u_t} = g_r \qquad r = 1, 2, \ldots, n \ . \tag{8a}$$

The *existence* of a *positive* solution $u_r = \bar{u}_r > 0$ is therefore established for the "irreducible" case.

However, in order to show also the *uniqueness* of this solution, we make use of an easily derived auxiliary formula. By summation of equations (8a) for $r = 1, 2, \ldots, p$, where $1 \leqq p \leqq n$, and upon noting the identity

$$k_{rt} \frac{u_r}{u_r + u_t} + k_{tr} \frac{u_t}{u_t + u_r} = k_{rt} = k_{tr} \ ,$$

we obtain the formula

$$\sum_{r=1}^{p} \sum_{t=p+1}^{n} k_{rt} \frac{u_r}{u_r + u_t} = \sum_{r=1}^{p} g_r - \sideset{}{'}\sum_{1 \leqq r, t \leqq p} k_{rt} = G_p - K_p = \gamma_p \geq 0 \ . \tag{15}$$

Here the right side expresses the excess γ_p of the total number of wins of the players $A_1, A_2, \ldots, A_p$ beyond the number K_p of games played among them; i.e., the number of games won by them against the *remaining* players A_s, which in the "irreducible" case, where a division into classes with $g_{rs} = 0$ is excluded, always comes out *positive*.

Suppose now that u_t and u_t' are two nontrivially different positive solutions of the set of homogeneous equations (8a), i.e., solutions that differ not only by a common multiplier ϱ. We may regard both u_t and u_t' as simultaneously normed to sum to 1,

$$\sum_{t=1}^{n} u_t = \sum_{t=1}^{n} u_t' - 1 \ , \tag{16}$$

and choose the sequence of the u_t in such a way that the first p of these quantities are *smaller*, but all the remaining u_t are larger than or equal to the corresponding u'_t; i.e.,

$$u_r < u'_r \text{ and } u_s \geq u'_s \text{ for } 1 \leq r \leq p < s \leq n. \tag{17}$$

It follows that, for every such combination r, s

$$\frac{u_r}{u_s} < \frac{u'_r}{u'_s} \tag{17a}$$

as well as

$$\frac{u_r}{u_r + u_s} < \frac{u'_r}{u'_r + u'_s} \tag{17b}$$

and hence also

$$\sum_{r=1}^{p} \sum_{s=p+1}^{n} k_{rs} \frac{u_r}{u_r + u_s} < \sum_{r=1}^{p} \sum_{s=p+1}^{n} k_{rs} \frac{u'_r}{u'_r + u'_s}. \tag{18}$$

This contradicts (15), provided even *one* of the numbers $k_{rs} > 0$, i.e., provided the players A_r have played at all against A_s, which is always so in the irreducible case, because of $k_{rs} \geq g_{rs} > 0$.

In the case considered, the *ratios* of the (positive) quantities u_t are therefore determined *uniquely* by the equations (8a). Also *every* positive solution of the set of equations, as the *only* solution of the *irreducible* problem, must yield the maximum $\bar{w}$ of the function $\Phi(u)$ defined by (3).

With the help of formula (15) we can derive yet another important property of our set of solutions. We consider again two different positive solutions u_t and u'_t of the equations (8a), which we regard also as normed according to (16). However, the u_t and u'_t may correspond to two *different* distributions of the number of wins g_t and g'_t, i.e., to different tournament results. Again, we suppose the sequence of the u_t chosen in such a way that the relations (17), (17a), and (17b) hold for $r \leq p < s$.[6] Since $k_{rt} = k'_{rt}$, according to our assumption that the number of games played is to be the same, we again obtain the inequality (18) and because of (15) also

$$G_p = g_1 + g_2 + \cdots + g_p < g'_1 + g'_2 + \cdots + g'_p = G'_p. \tag{19}$$

Thus, for every transition from a tournament result g_{rt} to another with g'_{rt}, the total number of wins grows for that group of players A_r, whose normed playing strength increases. Suppose now that in such a transition g_q is the *only* increasing g_t, so that $g_q < g'_q$, but otherwise $g_t \geq g'_t$. Then g_q must necessarily appear in the partial sum G_p; i.e., $q \leq p$ and $u_q < u'_q$.

---

[6] We can exclude the case where all $u_t = u'_t$, since in view of (8a), we would have $g_t = g'_t$ for all t, contrary to assumption.

Thus the relative (normed) playing strength of a player A_r will necessarily be increased by his winning rather than losing a single game, other results remaining unchanged.[7]

3 Omitted

4 Discussion of the Solution for Regular Tournaments

In the important case in practice, which we will designate as "regular," where all

$$k_{rs} = k$$

are equal, i.e., where each player has played the same number of games against every other player, eq. (8) becomes

$$g_r^* \equiv \frac{g_r}{k} = u_r \sum_t' \frac{1}{u_r + u_t} \qquad r = 1, \ldots, n \geqq 2 . \tag{8b}$$

If in the sequence $u_1, \ldots, u_n$ two values are equal: $u_r = u_s$, then it follows that

$$g_r = g_s .$$

But if

$$u_r < u_s \tag{26}$$

then also

$$\frac{1}{u_r + u_t} > \frac{1}{u_s + u_t} \tag{27}$$

for all $r \neq t \neq s$ and

$$\frac{u_r}{u_r + u_t} < \frac{u_s}{u_s + u_t} . \tag{28}$$

Thus

$$g_r < g_s ,$$

but, provided $n > 2$,

$$\frac{g_r}{u_r} > \frac{g_s}{u_s}$$

or

$$\frac{u_r}{u_s} < \frac{g_r}{g_s} < 1 . \tag{29}$$

Conversely, if $g_r \leqq g_s$, it follows again that $u_r \leqq u_s$.

---

[7] The importance of this requirement for a usable evaluation method is stressed by E. Landau, loc. cit. §3, where he also shows that the requirement is not always satisfied in his attempts so far to solve this problem.

We therefore have the

Theorem. *For a tournament, in which each player plays the same number of games with every other player, the respective playing strengths u_t are — in the irreducible case — in the same order as the numbers of wins g_t, but for $n > 2$ are relatively more widely spaced.*

In our case of $k_{rs} = k$, if the g_r (and thereby the u_r) are arranged in increasing order of magnitude

$$1 \leqq g_1 \leqq g_2 \leqq \cdots \leqq g_n ,$$

then every pth partial sum

$$g_1 + g_2 + \cdots + g_p \qquad (p = 1, 2, \ldots, n - 1)$$

will, by (15), turn out larger than the corresponding partial sum K_p:

$$G_p = g_1 + g_2 + \cdots + g_p > K_p = \sideset{}{'}\sum_{1 \leqq r, t \leqq p} k_{rt} = \frac{kp(p-1)}{2} ; \qquad (30)$$

only for $p = n$ do we have

$$g_1 + g_2 + \cdots + g_n = kn(n-1)/2 = N . \qquad (7a)$$

By partitioning the integer N into n components we obtain therefore in the above case all the possible tournament outcomes as expressed by the number of wins g_t.

Thus, e.g., in the case $k = 1$, $n = 2$, $N = 1 = 0 + 1$, we have only one possible partition

$$g_1 = 0 , \quad g_2 = 1 ,$$

which belongs to the "reducible" case and corresponds to the evaluation $u_1/u_2 = 0$.

For $k = 2$, $n = 2$, $N = 2 = 0 + 2 = 1 + 1$ we have also the irreducible partition $g_1 = g_2 = 1$, $u_1 = u_2$.

For $k = 3$, $n = 2$, $N = 3 = 1 + 2$ the irreducible partition $g_1 = 1$, $g_2 = 2$ arises. Correspondingly u_1 and u_2 are determined by equations (8)

$$\frac{3u_1}{u_1 + u_2} = 1 , \qquad \frac{3u_2}{u_1 + u_2} = 2 ,$$

with the result $u_1 : u_2 = 1 : 2$.

In general, for $n = 2$ we have $N = k = g_1 + g_2$,

$$g_1 = k\frac{u_1}{u_1 + u_2} , \qquad g_2 = k\frac{u_2}{u_1 + u_2} ,$$

so that $u_1 : u_2 = g_1 : g_2$; i.e., the calculated playing strengths behave like the numbers of wins, whereas for $n > 2$ eq. (29) always gives for $r < s$,

$$\frac{u_r}{u_s} < \frac{g_r}{g_s} < 1 \ .$$

For $n = 3$, $k = 1$, $N = 3 = g_1 + g_2 + g_3 = 1 + 1 + 1$ we have only a single irreducible outcome

$$g_1 = g_2 = g_3 = 1 \ , \quad u_1 = u_2 = u_3 \ ,$$

whereas for the case $n = 3$, $k = 2$, $N = 6$ there are three partitions

$$6 = 1 + 1 + 4 = 1 + 2 + 3 = 2 + 2 + 2$$

and hence three different sets of estimates. However, the first of these is *reducible*, since $g_1 + g_2 = 2 = 2.2(2 - 1)/2$, corresponding to the matrix $g_{12} = g_{21} = 1$, $g_{13} = g_{23} = 0$, $g_{31} = g_{32} = 2$. It would also arise if the players had met each other only once, and the first two had drawn and lost against the third. Thus, apart from the trivial outcome $g_1 = g_2 = g_3$, $u_1 = u_2 = u_3$, there remains only *one* irreducible case $g_1 = 1$, $g_2 = 2$, $g_3 = 3$. If we set

$$\frac{u_1}{u_3} = x \ , \quad \frac{u_2}{u_3} = y \ ,$$

we obtain the equations for x and y

$$\frac{x}{x + y} + \frac{x}{x + 1} = \frac{1}{2} \ , \quad \frac{y}{x + y} + \frac{y}{y + 1} = 1 \ ,$$

and hence

$$x = y^2 \ , \quad 1 = y + y^2 + 3y^3 \ .$$

The cubic equation has as unique positive solution

$$y = 0.46940 \ , \quad x = y^2 = 0.22034 \ .$$

Normed to sum to 1, the resulting playing strengths are

$u_1 = 0.1304$	$g_1 = 1 =$		1	$+$	0	$=$		0	$+$	1
$u_2 = 0.2778$	$g_2 = 2 =$	1		$+$	1	$= 2$			$+$	0
$u_3 = 0.5918$	$g_3 = 3 =$	2	$+ 1$			$= 1$	$+$	2		.

In particular, this set of estimates arises if each of the three players has played only *one* game with every other player and

1. A_1 has drawn against A_2 and lost against A_3; A_2 has drawn against A_3, or
2. A_1 has lost against A_2 and drawn against A_3; A_2 has lost against A_3.

For $n = 4$, $k = 1$ we again have only a single "irreducible" outcome, corresponding to the partition

$$N = 6 = 1 + 1 + 2 + 2 \; ,$$

with the result

$$u_1 : u_2 : u_3 : u_4 = 1 : 1 : 3 : 3 \; .$$

In contrast, for $n = 4$, $k = 2$ there arise already 7 irreducible partitions

$$
\begin{aligned}
N = 12 \; &= \; 1 + 2 + 4 + 5 \\
&= \; 1 + 3 + 3 + 5 \\
&= \; 1 + 3 + 4 + 4 \\
&= \; 2 + 2 + 3 + 5 \\
&= \; 2 + 2 + 4 + 4 \\
&= \; 2 + 3 + 3 + 4 \\
&= \; 3 + 3 + 3 + 3 \; .
\end{aligned}
$$

With the exception of the last (with the trivial solution $u_1 = u_2 = u_3 = u_4$) these lead to equations of the third or higher degree. E.g., the second yields

$$\frac{u_1}{u_4} = x = y^2 \; , \quad \frac{u_2}{u_4} = \frac{u_3}{u_4} = y \; , \quad 5y^3 + y^2 + 3y - 1 = 0 \; .$$

For $n = 5$, $k = 1$ we obtain 3 irreducible partitions

$$
\begin{aligned}
N = 10 \; &= \; 1 + 1 + 2 + 3 + 3 \\
&= \; 1 + 2 + 2 + 2 + 3 \\
&= \; 2 + 2 + 2 + 2 + 2 \; ,
\end{aligned}
$$

the first two again leading to cubic equations.

The fact that for $k_{rs} = k$ the playing strengths u_t are ordered in the same way as the corresponding number of wins g_t holds also in the *reducible* case. For if we have a division of the players into classes A_r and A_s, such that $g_{rs} = 0$, $1 \leqq r \leqq p < s \leqq N$, then for every g_r and every g_s

$$g_r \leqq k(p-1) < kp \leqq g_s \; . \tag{31}$$

This is so since a player A_r of the first class wins at most k games against each of $p - 1$ players in the same class and, on the other hand, each player of the second class wins at least kp games against all in the first class. Each number of wins in the lower class is therefore smaller than any in the upper class. At the same time each playing strength u_r belonging to the lower class vanishes against any u_s belonging to the upper class. In the case considered here it follows that if the g_r are arranged in increasing order, then every division into classes occurs between two successive g_p, g_{p+1} for all p for which

$$G_p = g_1 + g_2 + \cdots + g_p = k\frac{p(p-1)}{2} = k + 2k + \cdots + (p-1)k . \qquad (32)$$

Otherwise, because of (30), we always have

$$G_p = g_1 + g_2 + \cdots + g_p > k\frac{p(p-1)}{2} .$$

It is therefore sufficient to compare the successive partial sums of the ordered number of wins $g_1 \leqq g_2 \leqq g_3 \leqq \cdots \leqq g_n$ with the corresponding partial sums of the sequence $0, k, 2k, \ldots, (n-1)k$ in order to obtain immediately the complete decomposition into "prime tournaments."

5   The Numerical Solution of the Equations Through Successive Approximation

For larger values of n, the calculation through algebraic elimination is no longer feasible. An approximation procedure can then be recommended that, as we will show, always leads to the desired result in the irreducible case.

Equations (8) for the calculation of the relative playing strengths $\bar{u}_r$, now simply denoted by u_r, may be written in the form

$$\frac{g_r}{u_r} = {\sum_t}' \frac{k_{rt}}{u_r + u_t} \qquad \begin{pmatrix} r,t = 1,2,\ldots,n \\ t \neq r \end{pmatrix} . \qquad (8)'$$

It is therefore natural successively to improve given (positive) approximate values x_r of the u_r by setting

$$\frac{g_r}{x'_r} = {\sum_t}' \frac{k_{rt}}{x_r + x_t} = \frac{g_r}{f_r(x)} \ , \quad \text{i.e.,} \quad x'_r = f_r(x) \qquad (33)$$

and to proceed in turn from a set (x_r) to further sets $(x'_r), (x''_r), \ldots, (x_r^{(i)})$ which, if they converge, will yield the solution sought. For if, say,

$$\bar{x}_r = \lim_{i \to \infty} x_r^{(i)} \qquad (r = 1,\ldots,n) ,$$

then, analogously to (33), we obtain

$$\frac{g_r}{x_r^{(i+1)}} - {\sum_t}' \frac{k_{rt}}{x_r^{(i)} + x_t^{(i)}} \qquad (33)_t$$

and, proceeding to the limit,

$$\frac{g_r}{\bar{x}_r} = \sum_t \frac{k_{rt}}{\bar{x}_r + \bar{x}_t} \; ; \tag{34}$$

that is, the limiting values $\bar{x}_r$ form a set of solutions of (8).

In order to prove the actual convergence in the irreducible case of our problem, we treat a more general iteration problem, the assumptions for which are easily shown to apply in our case.

Theorem. *Consider a transformation of n variables x_r, in the form*

$$x'_r = f_r(x) = f_r(x_1, x_2, \ldots, x_n), \qquad (r = 1, 2, \ldots, n) \tag{35}$$

where the (real) functions f_r have the following properties:

1. The functions f_r are homogeneous of dimension 1; i.e., for every (positive) λ, we always have

$$f_r(\lambda x) = \lambda f_r(x) \; . \tag{36}$$

2. The functions are "positively restricted"; i.e., their values lie between finite positive bounds (different from zero) for every region of variation for which all the variables x_r themselves lie between finite positive bounds.

3. The functions f_r are continuously differentiable and all their partial derivatives $f_{rs}(x) = \frac{\partial}{\partial x_s} f_r(x)$ are also "positively restricted" functions, insofar as the variables x_s "actually occur"; i.e., the derivatives f_{rs} do not vanish identically.

4. Each of the functions f_r "actually" contains the variable x_r corresponding to it, so that $f_{rr} \neq 0$ in every positive region of the variables. Also for every division of the numbers $1, 2, \ldots, n$ into two classes (r, s), there always exists at least one pair of numbers r, s with r in one class and s in the other, for which $f_{rs} \neq 0$.

5. The equations

$$u_r = f_r(u) \qquad (r = 1, 2, \ldots, n) \tag{37}$$

have at least one positive solution $u_r > 0$.

Then, for every positive set of values $x_r > 0$, the successively formed values

$$x'_r = f_r(x), \; x''_r = f_r(x'), \ldots, x_r^{(i+1)} = f_r(x^{(i)}), \ldots \tag{35}_i$$

tend, as i increases, to positive limiting values

$$\bar{x}_r = \lim_{i \to \infty} x_r^{(i)} \qquad (r = 1, 2, \ldots, n) \; . \tag{38}$$

These satisfy the set of equations (37) and differ from the stated solutions u_r at most by a common positive factor:

$$\bar{x}_r = \lambda^* u_r \; , \tag{39}$$

which ensures at the same time the uniqueness of the positive solution.

Before proving this general theorem, we will first check that the conditions on the functions f_r are indeed satisfied in the special case treated by us.

1. The homogeneity follows immediately from the form of equations (8).

2. From Section 1 all $g_r > 0$ in the "irreducible" case, and for every r at least one $k_{rt} > 0$. The functions f_r are therefore nonvanishing rational functions with all coefficients *positive*. But such functions are always "positively restricted," since, as is easily seen, the sum, the product, and the ratio of "positively restricted" variables possess this same property. Quite generally, all positively restricted functions of positively restricted functions are themselves positively restricted.

3. By partial differentiation we have from (33)

$$g_r \frac{\partial f_r(x)}{\partial x_r} = f_r(x)^2 \sum_t' \frac{k_{rt}}{(x_r + x_t)^2} \tag{40}$$

and

$$g_r \frac{\partial f_r(x)}{\partial x_s} = f_r(x)^2 \frac{k_{rs}}{(x_r + x_t)^2} \quad \text{for} \quad s \neq r . \tag{41}$$

Thus all $f_{rs}(x)$, insofar as they do not vanish identically, are here also continuous and "positively restricted" functions.

4. From the last formulae it follows directly that always $f_{rr}(x) > 0$ and that $f_{rs}(x) = 0$ only if $k_{rs} = 0$, when the rth player has not played the sth at all. If there were a division into classes A_r and A_s, for which all k_{rs} are zero, then all g_{rs} would also be zero, which is excluded in the "irreducible" case.

5. The *existence* of a positive solution of the set of equations is assured in the irreducible case by our arguments in Section 1 regarding the existence of the maximum of Φ according to Weierstrass's theorem. On the other hand, the *uniqueness* of the solution does *not* have to be assumed here. Rather, independently of our proof in Section 2, this will be proved as a special consequence of the general theorem.

Proof of the Convergence Theorem. For $t = 1, 2, \ldots, n$, let $x_t > 0$ be an arbitrary set of positive values and $u_t > 0$ a positive solution of equations (37). If we denote by λ the smallest and by μ the largest ratio in the sequence x_t/u_t,

$$\lambda u_t \leqq x_t \leqq \mu u_t , \qquad (t = 1, 2, \ldots, n) \tag{42}$$

it follows from 3. that

$$\lambda f_r(u) \leqq f_r(x) \leqq \mu f_r(u) , \qquad (r = 1, 2, \ldots, n) \tag{43}$$

since the functions f_r are monotone. It also follows from (37) that

$$\lambda u_r \leqq x_r' \leqq \mu u_r$$

as well as that

$$\lambda u_t \leq x_t^{(i)} \leq \mu u_t \qquad (t = 1, 2, \ldots, n) \tag{42}_i$$

for all subsequent approximations $x_t^{(i)}$.

Likewise, if for every i, $\lambda^{(i)}$ again denotes the smallest and $\mu^{(i)}$ the largest ratio $x_t^{(i)}/u_t$, we have also

$$\lambda^{(i)} u_t \leq x_t^{(i)} \leq \mu^{(i)} u_t$$

and

$$\lambda^{(i)} u_t \leq x_t^{(i+1)} \leq \mu^{(i)} u_t .$$

Hence

$$\lambda^{(i)} \leq \lambda^{(i+1)} \leq \mu^{(i+1)} \leq \mu^{(i)} ;$$

i.e., the values $\lambda^{(i)}$ form a never decreasing and the $\mu^{(i)}$ a never increasing sequence, so that the intervals $(\lambda^{(i)}, \mu^{(i)})$ lie successively inside one another and all are contained in the overall interval (λ, μ). Thus the $\lambda^{(i)}$ possess an *upper* bound $\bar{\lambda}$, the $\mu^{(i)}$ a *lower* bound $\bar{\mu}$, and we have

$$\lambda \leq \lambda^{(i)} \leq \bar{\lambda} \leq \bar{\mu} \leq \mu^{(i)} \leq \mu \quad \text{for} \quad i = 1, 2, 3, \ldots \tag{43}$$

as well as

$$\bar{\lambda} = \lim_{i \to \infty} \lambda^{(i)} , \quad \bar{\mu} = \lim_{i \to \infty} \mu^{(i)} . \tag{44}$$

If we can now show that the two limiting values coincide:

$$\bar{\lambda} = \bar{\mu} = \varrho , \tag{45}$$

then we also have

$$\lim_{i \to \infty} x_t^{(i)} = \varrho u_t , \qquad (t = 1, 2, \ldots, n) \tag{46}$$

as is to be shown.

However, to establish this, we must distinguish two cases:

Case I. All $f_{rs} > 0$; i.e., in the tournament problem k_{rs} is positive for all $r \neq s$: every player has met every other player. Then, by 3. all derivatives f_{rs} are "positively restricted" and have positive lower bounds

$$f_{rs}(x) \geq \gamma_{rs} > 0 \tag{47}$$

for all values x_t (including also all $x_t^{(i)}$ by $(42)_i$) that satisfy the conditions

$$\lambda u_t \leq x_t \leq \mu u_t \qquad (t = 1, 2, \ldots, n) . \tag{42}$$

But by the mean value theorem we have, in view of (37),

$$x_r^{(i+1)} - \lambda^{(i)} u_r = f_r(x^{(i)}) - f_r(\lambda^{(i)} u)$$

$$= \sum_{t-1}^{n} f_{rt}(\tilde{x})(x_t^{(i)} - \lambda^{(i)} u_t) , \qquad (48)$$

where

$$\tilde{x}_t = (1 - \vartheta_r)\lambda^{(i)} u_t + \vartheta_r x_t^{(i)} \qquad (0 \leq \vartheta_r \leq 1; \ t = 1, 2, \dots, n)$$

denotes a point on the line connecting the points $(x_t^{(i)})$ and $(\lambda^{(i)} u_t)$, which likewise satisfies the conditions (42). Hence by (47), for every $s \leq n$,

$$x_r^{(i+1)} - \lambda^{(i)} u_r \geq \gamma_{rs}(x_s^{(i)} - \lambda^{(i)} u_s) . \qquad (49)$$

In particular, this holds also for the maximum and minimum values, respectively, of the ratios $[x_t^{(i)}/u_t]$, where

$$x_r^{(i+1)} = \lambda^{(i+1)} u_r , \quad x_s^{(i)} = \mu^{(i)} u_s ,$$

so that we obtain

$$\lambda^{(i+1)} - \lambda^{(i)} \geq \gamma_{rs}\frac{u_s}{u_r}(\mu^{(i)} - \lambda^{(i)}) \geq \kappa(\mu^{(i)} - \lambda^{(i)}) , \qquad (50)$$

where κ denotes the smallest of the positive values $\kappa_{rs} = \gamma_{rs}u_s/u_r$. It follows that

$$\mu^{(i)} - \lambda^{(i)} \leq \frac{1}{\kappa}(\lambda^{(i+1)} - \lambda^{(i)})$$

and

$$\bar{\mu} - \bar{\lambda} = \lim_{i \to \infty} (\mu^{(i)} - \lambda^{(i)}) = 0 , \qquad (51)$$

as was to be proved.

In order to estimate also the rate of convergence, we make use of a second formula, derived quite analogously to (50) from (47). We have

$$\mu^{(i)} u_r - x_r^{(i+1)} = f_r(\mu^{(i)} u) - f_r(x^{(i)})$$

$$= \sum_{t=1}^{n} f_{rt}(\tilde{x})(\mu^{(i)} u_t - x_t^{(i)})$$

$$\geq \gamma_{rs}(\mu^{(i)} u_s - x_s^{(i)})$$

for arbitrary r, s. Thus for

$$x_r^{(i+1)} = \mu^{(i+1)} u_r , \quad x_s^{(i)} = \lambda^{(i)} u_s$$

this gives

$$\mu^{(i)} - \mu^{(i+1)} \geq \kappa(\mu^{(i)} - \lambda^{(i)}) . \qquad (52)$$

By addition of (50) and (52) we then obtain

$$\mu^{(i)} - \lambda^{(i)} - (\mu^{(i+1)} - \lambda^{(i+1)}) \geq 2\kappa(\mu^{(i)} - \lambda^{(i)}) \ .$$

Thus

$$\mu^{(i+1)} - \lambda^{(i+1)} \leqq (1 - 2\kappa)(\mu^{(i)} - \lambda^{(i)}) \tag{53}$$

so that

$$\mu^{(i)} - \lambda^{(i)} \leqq (\mu - \lambda)(1 - 2\kappa)^i \ . \tag{54}$$

The approximations therefore converge at least as fast as a geometric series.

Case II. Not all f_{rs} are unequal to zero, not all f_r "actually" contain all variables, but property 4 holds. This case, as we will show, can be reduced to Case I by iterating the transformations (35) and setting

$$
\begin{aligned}
f_r^{(1)}(x) &\equiv f_r(f) \equiv f_r\Big(f_1(x), f_2(x), \dots, f_n(x)\Big) \\
f_r^{(2)}(x) &\equiv f_r(f^{(1)}) \equiv f_r\Big(f_1^{(1)}(x), f_2^{(1)}(x), \dots, f_n^{(1)}(x)\Big) \\
&\cdots\cdots\cdots\cdots\cdots\cdots\cdots\cdots\cdots\cdots \\
f_r^{(i+1)}(x) &\equiv f_r(f^{(i)}) \equiv f_r\Big(f_1^{(i)}(x), \dots, f_n^{(i)}(x)\Big) \\
&(r = 1, 2, \dots, n; \ i = 1, 2, \dots) \ .
\end{aligned}
\tag{55}
$$

These functions $f_r^{(i)}(x)$, arising through iteration, are easily seen to possess the same properties 1–3 as do the original $f_r(x)$, since these properties always carry over to the functions formed. In particular, for every such function F in every positively restricted region, we have also

$$\frac{\partial}{\partial x_r} F(f_1, f_2, \dots) = \sum_t \frac{\partial F}{\partial f_t} \frac{\partial f_t}{\partial x_r} \geqq \frac{\partial F}{\partial f_s} \frac{\partial f_s}{\partial x_r} > 0 \ , \tag{56}$$

if even for *one* s the derivative of F w.r.t. f_s as well as the derivative of f_s w.r.t. x_r do not vanish. From this it follows directly that with $f_{rr} \neq 0$ (in view of 4.) we have also

$$f_{rr}^{(1)}(x) > 0, \ f_{rr}^{(2)}(x) > 0, \dots, f_{rr}^{(i)}(x) > 0 \ ,$$

so that all $f_r^{(i)}$ are *monotone* functions of x_r.

Suppose now that the function $F(x) = F(x_1, x_2, \dots, x_n)$, taken from the sequence of functions f_t, "actually" contains the variables denoted by x_r in the sequence $x_1 \dots x_n$, so that all $\partial F / \partial x_r > 0$, whereas for all the remaining variables (here denoted by x_s) we always have $\partial F / \partial x_s = 0$. Then by 4. there exists a pair of values r, s for which $f_{rs} \neq 0$, so that

$$\frac{\partial}{\partial x_r} F(f) > 0 \text{ for all } r \text{ for which } \frac{\partial F(x)}{\partial x_r} > 0 \tag{57}$$

as well as

$$\frac{\partial}{\partial x_s} F(f) \geq \frac{\partial F}{\partial f_r} \frac{\partial f_r}{\partial x_s} > 0 ; \tag{58}$$

i.e., the function $F(f) = F^{(1)}(x)$ actually contains *all* x_r actually contained in $F(x)$, and *in addition* at least one more variable x_s.

Thus, in the sequence

$$F(x), \ F^{(1)}(x) = F(f), \ F^{(2)}(x) = F^{(1)}(f), \ldots, F^{(i+1)}(x) = F^{(i)}(f) ,$$

already $F^{(n-1)}(x)$ contains all n variables $x_1, \ldots, x_n$. In particular, there exists also a number $h \leq n - 1$ with the property that the n functions

$$f_1^{(h)}(x), f_2^{(h)}(x), \ldots, f_n^{(h)}(x) ,$$

which arise through h-fold iteration of transformation (35), actually contain all the variables $x_1, \ldots, x_n$. These functions $f_r^{(h)}$ satisfy all of the conditions 1 to 5, since evidently for every set of positive solutions (u) of (37) we also have

$$u_r = f_r^{(h)}(u) \qquad (r = 1, 2, \ldots, n) . \tag{37$_h$}$$

Thus all the assumptions needed for convergence in Case I are met. For an arbitrary set (x) of initial values, the iterates[9]

$$x_r^{(h)} = f_r^{(h)}(x), \ x_r^{(2h)} = f_r^{(2h)}(x), \ldots, x_r^{(ih)} = f_r^{(ih)}(x)$$

will therefore tend, with increasing i, to a solution of (37), since the corresponding differences $\mu^{(ih)} - \lambda^{(ih)}$ approach 0. So we have quite generally

$$\bar{\mu} - \bar{\lambda} = \lim_{j=\infty} (\mu^{(j)} - \lambda^{(j)}) = 0 ;$$

i.e., in the *general* case given by conditions 1 to 5 our procedure of successive approximation also leads to the numerical solution of the irreducible problem. The difference between Cases I and II becomes noticeable only in that for Case II the sequence of approximations $x^{(i)}$ occasionally results in

$$\mu^{(i+1)} - \lambda^{(i+1)} = \mu^{(i)} - \lambda^{(i)} ,$$

whereas, as follows from our argument regarding $f_r^{(h)}$, we must always have

$$\mu^{(i+h)} - \lambda^{(i+h)} < \mu^{(i)} - \lambda^{(i)} .$$

---

[9] We make use here of the validity of the associative law for the combination of transformations.

Our method of solution is therefore far from being confined only to "regular" tournaments (where $k_{rt} = k$, a constant), as are to the best of my knowledge all previous methods of tournament evaluation. In the "irreducible" case our method can also be applied without change to "broken-off" and "combined" tournaments (made up of several individual tournaments with partly identical players). In "reducible" cases also the method always furnishes a complete or partial determination of the relative playing strengths, which treats all available data equally and satisfies all sensible requirements.

In the practical execution of the computations one will, as a rule, begin with the initial values $x_1 = x_2 = \cdots = x_n = 1$ and, with the help of tables of reciprocals, determine successive approximations $x'_r, x''_r, \ldots$ according to (33). This process continues until each two successive values no longer differ significantly, so that the set of equations (8) is satisfied. However, first one must have checked the "irreducibility" of the tournament and, if necessary, undertaken the required decomposition into "prime tournaments." The latter is most conveniently done, in the "regular" case ($k_{rt} = k$), according to the procedure given at the end of Section 4, i.e., by arrangement of the g_r in increasing order and comparison of their partial sums G_r with the corresponding numbers $\frac{1}{2}kr(r - 1)$, according to (32).

I conclude with an example of the evaluation of a fairly large (regular) tournament. The short table below was constructed by the methods described, following the New York Masters Tournament of 1924 with $n = 11$ participants (including E. Lasker as first prize winner). The computed "playing strengths" u_r are normed to sum to 100, and γ_r denotes the difference $G_r - r(r - 1)$, since $k = 2$.

r	u_r	g_r	G_r	$r(r-1)$	γ_r
1	2.43	5	5	0	5
2	3.44	6.5	11.5	2	9.5
3	3.84	7	18.5	6	12.5
4	4.72	8	26.5	12	14.5
5	6.40	9.5	36	20	16
6	7.12	10	46	30	16
7	7.84	10.5	56.5	42	14.5
8	8.71	11	67.5	56	11.5
9	10.7	12	79.5	72	7.5
10	18.4	14.5	94	90	4
11	26.4	16	110	110	0

(Received May 28, 1928)

The Origin of Confidence Limits

Comments on Fisher (1930)

1 Introduction

In the history of ideas it is frequently possible, with the advantage of hindsight, to discern earlier examples of new concepts. Their later appreciation often relies on the clarification of thought accompanying the introduction of terms which distinguish previously confused concepts. In statistics a notable example is provided by the separation of probability and likelihood.

Here we present the first paper (Fisher, 1930) which clearly identifies the *confidence concept*, though not under that name. Limiting the discussion to the case of a single parameter, as does the paper, by the *confidence concept* we mean the idea that a probability statement may be made about an unknown parameter (such as limits between which it lies, or a value which it exceeds) which will be correct under repeated sampling from the same population.

The implicit use of the confidence concept may be traced back to the work of Laplace and Gauss at the beginning of the nineteenth century, especially Gauss (1816) which we reproduce in translation in this volume (see also Lehmann, 1958 and Hald, 1998). Laplace and Gauss were quite clear that it had a "repeated sampling" rationale as opposed to a Bayesian one (Gauss even using betting language), but the confidence limits found were usually approximate, as in the cases of a binomial parameter and the mean of a normal distribution with estimated variance, so that the logic of the procedure was obscured. The practice of attaching a "probable error" to an estimate thus became commonplace without a full understanding of its rationale (see, however, Cournot, 1843, p. 186, and Hald's (2000) comments on the work of Pizzetti). The importance of Fisher's 1930 paper lies in its clear exposition of the underlying logic, and the influence the paper had on both J. Neyman and E.S. Pearson.

An interesting example from before 1930 was noticed much later by Fisher (1939) himself: Student (1908), in his famous paper on the *t*-distribution, had casually remarked: "... if two observations have been made and we have no other information, it is an even chance that the mean of the (normal) population will lie between them." In modern language, the two observations themselves form a 50% *confidence interval* for the unknown mean. It may not be a very good one according to some auxiliary criterion, but that is not our present concern; all that matters is that if on each occasion on which we draw the sample of two at random we make the statement "the mean lies in the interval," half the time we will be correct.

Student's example is remarkable for being exact, but although his statement and other similar ones from his paper (see especially Section VIII: Explanation of Tables, noticed by Welch, 1939) are couched in language which makes them indistinguishable from confidence statements, it has to be accepted that they are no more than an ellipsis of the type which had long been customary when attaching limits to parameters. The advance made by Fisher in 1930 was his recognition that such statements could, in certain circumstances, be justified as exact statements of probability with a hitherto-unclarified logical basis. These he called statements of *fiducial probability*.

2   Fisher's 1930 Paper

The first half of Fisher's paper is a critique of inverse probability or, as we should now say, of Bayesian estimation theory. He uses a favorite argument of his concerning the effects of transformation, first seen in Fisher (1912). He points out that if a uniform prior distribution is assumed for a parameter, and the value which maximizes the posterior probability distribution (the mode) is taken as its estimate, the result is invariant to 1:1 transformations of the parameter. This, he argues, is an illusory property: the arbitrary nature of the first assumption, the uniform prior, serves only to cancel out the arbitrary nature of the second, the choice of the mode as the estimate. The method of maximum likelihood makes neither assumption, and Fisher refers to the "historical accident" that Gauss developed his analogous method of estimation in terms of inverse probability.

The second half of the paper gives a straightforward account (it now seems) of how to obtain exact statements of fiducial probability about the maximum likelihood estimate of a single parameter by an appropriate interpretation of the significance levels of the associated test of significance. Fisher uses the correlation coefficient as a numerical example, giving a table, and concludes the paper with a clear analysis of the difference between a statement of fiducial probability and one of inverse probability. In retrospect it is unfortunate that Fisher wrote of a fiducial *distribution*, for it gave the false impression that fiducial probabilities could be manipulated like any other probabilities, whereas in fact all that can be claimed is that a probability statement about a parameter based on its fiducial "distribution" has the same inductive content *as if* the parameter had been drawn at random from a conventional probability distribution of identical form.

The key concept is that of the confidence property (so named by Neyman, 1934). In repeated applications of the procedure the derived probabilities will be verifiable repeated-sampling probabilities, and statements about the parameter based on its fiducial probability distribution will be true with the assigned probability.

The title of the paper, "Inverse probability," which now seems a little misleading, is explained by the fact that in 1930 the phrase would not necessarily have been taken to refer exclusively to the Bayesian method (which in the paper Fisher calls "inverse probability strictly speaking") but to the general problem of arguing "inversely" from sample to parameter (see Edwards, 1997).

3   Reception of Fisher's 1930 Paper

In discussing the reception of Fisher's description of how exact probability statements about an unknown parameter can be made independently of a Bayesian formulation we do not comment on the subsequent history of fiducial theory and of the divergence between Fisher (who sought to understand the conditions which made fiducial statements valid inferences) and Neyman (who denied the possibility of such inferences and accepted only a behavioristic interpretation of confidence intervals). Reference may be made to Edwards (1995) and the books and papers cited therein, especially Zabell (1992). Here we concentrate solely on the confidence concept itself.

Neyman's first publication on confidence intervals was in 1934. The section of his paper in which he introduced and named them is headed "The Theory of Probabilities *a posteriori* and the work of R.A. Fisher." Referring to Fisher's papers, including Fisher (1930), Neyman wrote:

> *The possibility of solving the problems of statistical estimation independently from any knowledge of the* a priori *probability laws, discovered by R.A. Fisher, makes it superfluous to make any appeals to the Bayes' theorem.*

In a footnote he added:

> *The above-mentioned problems of confidence intervals are considered by R.A. Fisher as something like an additional chapter to [his] Theory of Estimation, being perhaps of minor importance. However, I do not agree in this respect with Professor Fisher. I am inclined to think that the importance of his achievements in the two fields is in a relation which is inverse to what he thinks himself. The solution of the problem which I described as the problem of confidence intervals has been sought by the greatest minds since the work of Bayes 150 years ago.*

Neyman's paper is ambivalent about the role of the Bayesian prior in the analysis. Whereas Fisher had unequivocally rejected a Bayesian framework and had expounded the difference between a fiducial probability and a posterior probability, Neyman started with the assumption of the existence of an unknown prior distribution and then discovered its irrelevance, which Fisher had already explained. Ironically, in Neyman's later "received" theory of confidence intervals his treatment involved sampling repeatedly from the *same* population, so that there was no question of a prior distribution, whereas Fisher had never imposed such a restriction. But when, in his definitive paper of 1937, Neyman abandoned all mention of a prior distribution and presented the confidence concept independently of Bayesian considerations, as Fisher had originally done, he also abandoned his reference to Fisher (1930), an omission which Bartlett "was always rather surprised to find" (Kendall et al., 1982).

According to Neyman's biographer (Reid, 1982) his 1934 paper was "an English version of the pamphlet already published in Polish [in 1933]." During the summer of 1933 he had become familiar with Fisher (1930) and he added the warm acknowledgment of it which his 1934 paper contains. "He cheerfully recognized Fisher's priority for a theory of interval estimation independent of probabilities a priori" and "He planned to publish what he wrote 'with compliments to Fisher'."

But from 1937 on, the growing appreciation of the differences in purpose between the fiducial theory (a theory of inference) and the confidence theory (a theory of behavior) led Neyman to underplay the fiducial origins of the basic confidence concept, claiming that "Fisher's method of approach was entirely different from the author's" (Neyman, 1941). He recounted that he had in fact developed the confidence concept "about 1930" and had incorporated it in his lectures, resulting in an actual application being published in Poland in 1932 (Pytkowski, 1932). Neyman gave further details in 1970, recalling that his original development had been *"quasi* Bayesian" and that Pytkowski had been his student. He explained that because "the specific solutions of several particular problems coincided" in his and Fisher's developments "in 1934 I recognized Fisher's priority for the idea that interval estimation is possible without any reference to Bayes' theorem and with the solution being independent from probabilities *a priori.*" He attributed to another student, Churchill Eisenhart, who attended his London lectures in 1935, the impetus for removing all notion of the parameter under estimation being a random variable.

By 1950 confidence intervals had become the received theory of parameter estimation, and in his book *First Course in Probability and Statistics* Neyman (1950) dated it from his 1934 paper. In a lecture at a history of statistics meeting in 1974 he said "A novel form of the problem of estimation was noticed by the present author, that of estimation 'by interval' .... Whereas the first publication regarding the confidence intervals goes back to 1930 (29), the definitive theory appeared ... in 1937" (Neyman, 1976). Reference (29) is, however, not to Fisher (1930), which is not mentioned, but to Pytkowski (1932). Again, in 1977, the historical account no longer mentioned Fisher (Neyman, 1977).

Nor was it only in the literature of confidence intervals that Fisher's contribution went unremarked. His biographer (Box, 1978) does not mention it, nor even does Rao (1992) in his more detailed survey "R.A. Fisher: The Founder of Modern Statistics." Neyman (1967), in his otherwise generous appreciation of Fisher published in *Science*, only remarked "In several earlier writings I have pointed out that certain of Fisher's conceptual developments, not mentioned here, are erroneous."

Late in life E.S.Pearson, Neyman's collaborator in the Neyman–Pearson theory of hypothesis testing, reminisced wistfully about how their conversations in the summer of 1926 might have led to confidence intervals "but the underlying philosophy had not yet surfaced" (Pearson, 1990). In a letter to W.S. Gossett in 1927 he "introduced in a crude visual way a method by which the problem of interval estimation might be tackled.... The method resembles the standard procedure of finding a 'confidence distribution' ...." (Pearson, 1990). Pearson also remarked on the importance of the new format for statistical tables pioneered by Fisher (1925) in *Statistical Methods for Research Workers*, which had enabled Pytkowski to compute a confidence interval without using interpolation. "Whether the availability of Fisher's type of tables in any way influenced Neyman in his putting forward his confidence interval ideas in his Warsaw lectures, I do not know." We do know, of course, that it was a primary influence for Fisher (1930), and that 1926 is also the year in which the first ideas about interval estimation were discussed between Fisher and his colleagues at Rothamsted (Edwards, 1995).

Clopper and Pearson (1934), introducing their "confidence or fiducial limits" for the case of the binomial parameter, wrote

> *The recent work of R.A. Fisher [1930, 1933] introducing the conception of the fiducial interval has made it possible under certain conditions to treat this problem of estimation in a simple yet powerful manner.*

and in a footnote they added "References to the discussion of these concepts in lectures may also be found in papers published by students of J. Neyman." However, they preferred Neyman's (1934) terminology to Fisher's:

> *In his development of the subject, R.A. Fisher has used the term "fiducial probability" to describe the chance that in the long run a correct prediction will be made of the limits within which the unknown parameter falls.... We are inclined ... to adopt the terminology suggested by J. Neyman, and to convey what is fundamentally the same notion by specifying the confidence coefficient associated with an interval.*

Additional, careful, comments on confidence intervals may be found in Bartlett's contribution to Neyman's Royal Society obituary notice (Kendall et al., 1982).

Fiducial theory and confidence theory went their separate ways. Eventually some understanding was reached as to the limits of applicability of fiducial theory as a theory of inference and the reasons why Fisher's early intuition that his fiducial probabilities could be manipulated was mistaken, but confidence theory, requiring only its secure mathematical foundations for its development, became for a while the received theory of (paradoxically) inference, since that has been its major use in spite of Neyman's protestations.

We leave the last word to Zabell (1992), who thought that "The original fiducial argument, for a single parameter, was virtually indistinguishable from the confidence approach of Neyman, ..."; "Fisher not only gave a clear and succinct statement of (what later came to be called) the confidence interval approach to parameter estimation, but (and this appears to be almost universally unappreciated) he also gave a general method for obtaining such estimates in the one-dimensional case."

References

Box, J.F. (1978). *R.A. Fisher, The Life of a Scientist.* Wiley, New York.

Clopper, C.J. and Pearson, E.S. (1934). The use of confidence or fiducial limits illustrated in the case of the binomial. *Biometrika,* **26**, 404–413.

Cournot, A.A. (1843). *Exposition de la théorie des chances et des probabilités.* Hachette, Paris.

Edwards, A.W.F. (1995). Fiducial inference and the fundamental theorem of natural selection. XVIIIth Fisher Memorial Lecture, London, 20th October 1994. *Biometrics,* **51**, 799–809.

Edwards, A.W.F. (1997). What did Fisher mean by "inverse probability" in 1912–1922? *Statistical Science,* **12**, 177–184.

Fisher, R.A. (1912). On an absolute criterion for fitting frequency curves. *Messenger Math.,* **41**, 155–160.

Fisher, R.A. (1925). *Statistical Methods for Research Workers.* Oliver & Boyd, Edinburgh.

Fisher, R.A. (1930). Inverse probability. *Proc. Camb. Phil. Soc.,* **26**, 528–535.

Fisher, R.A. (1933). The concepts of inverse probability and fiducial probability referring to unknown parameters. *Proc. Roy. Soc. Lond.,* A, **139**, 343–348.

Fisher, R.A. (1939). "Student." *Ann. Eugen.,* **9**, 1–9.

Hald, A. (1998). *A History of Mathematical Statistics from 1750 to 1930.* Wiley, New York.

Hald, A. (2000). Studies in the history of probability and statistics XLV. Pizzetti's contributions to the statistical analysis of normally distributed observations, 1891. *Biometrika,* **87**, 213–217.

Kendall, D.G., Bartlett, M.S., and Page, T.L. (1982). Jerzy Neyman. *Biographical Memoirs of Fellows of the Royal Society*, **28**, 379–412.

Lehmann, E.L. (1958). Some early instances of confidence statements. Technical Report to the Office of Naval Research ONR 5, Statistical Laboratory, University of California, Berkeley.

Neyman, J. (1934). On the two different aspects of the representative method. *J. Roy. Statist. Soc.*, **97**, 558–625 (with discussion).

Neyman, J. (1937). Outline of a theory of statistical estimation based on the classical theory of probability. *Phil. Trans. Roy. Soc.*, A, **236**, 333–380.

Neyman, J. (1941). Fiducial argument and the theory of confidence intervals. *Biometrika*, **32**, 128–150.

Neyman, J. (1950). *First Course in Probability and Statistics*. Holt, New York.

Neyman, J. (1967). R.A. Fisher (1890–1962): An appreciation. *Science*, **156**, 1456–1460.

Neyman, J. (1970). A glance at some of my personal experiences in the process of research. In: *Scientists at Work*, T. Dalenius, G. Karlsson, and S. Malmquist, eds. Almqvist and Wiksell, Stockholm, pp. 148–164.

Neyman, J. (1976). The emergence of mathematical statistics. In *On the History of Statistics and Probability*, D.B. Owen (ed.), Marcel Dekker, New York, pp. 149–193.

Neyman, J. (1977). Frequentist probability and frequentist statistics. *Synthese*, **36**, 97–131.

Pearson, E.S. (1990). *'Student': A Statistical Biography of W.S. Gosset*. Edited and augmented by R.L. Plackett with the assistance of G.A. Barnard. Clarendon, Oxford.

Pytkowski, W. (1932). The dependence of the income in small farms upon their area, the outlay and the capital invested in cows. (Polish, with English summaries.) *Biblioteka Pulawska 31*. Agri. Res. Inst., Pulawy.

Rao, C.R. (1992). R.A. Fisher: The founder of modern statistics. *Statist. Sci.*, **7**, 34 48.

Reid, C. (1982). *Neyman from Life*. Springer, New York.

'Student' (1908). The probable error of a mean. *Biometrika*, **6**, 1–25.

Welch, B.L. (1939). On confidence limits and sufficiency, with particular reference to parameters of location. *Ann. Math. Statist.*, **10**, 58–69.

Zabell, S.L. (1992). R.A. Fisher and the fiducial argument. *Statist. Sci.*, **7**, 369–387.

See also:

Aldrich, J. (2000). Fisher's "Inverse Probability" of 1930. *Intern. Statist. Rev.* **68**, 155–172.

Inverse Probability. By R. A. Fisher, Sc.D., F.R.S., Gonville and Caius College; Statistical Dept., Rothamsted Experimental Station.

[*Received* 23 July, *read* 28 July 1930.]

I know only one case in mathematics of a doctrine which has been accepted and developed by the most eminent men of their time, and is now perhaps accepted by men now living, which at the same time has appeared to a succession of sound writers to be fundamentally false and devoid of foundation. Yet that is quite exactly the position in respect of inverse probability. Bayes, who seems to have first attempted to apply the notion of probability, not only to effects in relation to their causes but also to causes in relation to their effects, invented a theory, and evidently doubted its soundness, for he did not publish it during his life. It was posthumously published by Price, who seems to have felt no doubt of its soundness. It and its applications must have made great headway during the next 20 years, for Laplace takes for granted in a highly generalised form what Bayes tentatively wished to postulate in a special case.

Before going over the formal mathematical relationships in terms of which any discussion of the subject must take place, there are two preliminary points which require emphasis. First, it is not to be lightly supposed that men of the mental calibre of Laplace and Gauss, not to mention later writers who have accepted their views, could fall into error on a question of prime theoretical importance, without an uncommonly good reason. The underlying mental cause is not to be confused with the various secondary errors into which one is naturally led in deriving a formal justification of a false position, such as for example Laplace's introduction into his definition of probability of the unelucidated phrase "equally *possible* cases" which, since we must be taken to know what cases are equally possible before we know that they are equally probable, can only lead to the doctrine, known as the "doctrine of insufficient reason," that cases are equally probable (to us) unless we have reason to think the contrary, and so reduces all probability to a subjective judgment. The underlying mental cause is, I suggest, not to be found in these philosophical entanglements, but in the fact that we learn by experience that science has its inductive processes, so that it is naturally thought that such inductions, being uncertain, must be expressible in terms of probability. In fact, the argument runs somewhat as follows: a number of useful but uncertain judgments can be expressed with exactitude in terms of probability; our judgments respecting causes or hypotheses are uncertain, therefore our rational attitude towards them is expressible in terms of probability. The assumption was almost a necessary one seeing that no other mathematical apparatus existed for dealing with uncertainties.

The second point is that the development of the subject has reduced the original question of the inverse argument in respect of probabilities to the position of one of a series of quite analogous questions; the hypothetical value, or parameter of the population under discussion, may be a probability, but it may equally be a correlation, or a regression, or, in genetical problems, a linkage value, or indeed any physical magnitude about which the observations may be expected to supply information. The introduction of quantitative variates, having continuous variation in place of simple frequencies as the observational basis, makes also a remarkable difference to the kind of inference which can be drawn.

It will be necessary to summarise some quite obvious properties of these continuous frequency distributions. The probability that a variate x should have a value in the range $x \pm \frac{1}{2} dx$ is expressed as a function of x in the form

$$df = \phi(x)\, dx.$$

The function depends of course on the particular population from which the value of x is regarded as a random sample, and specifies the distribution in that population. If in the specification of the population one or more parameters, $\theta_1, \theta_2, \theta_3, \ldots$ are introduced, we have

$$df = \phi(x, \theta_1, \theta_2, \theta_3, \ldots)\, dx,$$

where ϕ now specifies only the form of the population, the values of its parameters being represented by $\theta_1, \theta_2, \theta_3, \ldots$.

Knowing the distribution of the variate x, we also know the distribution of any function of x, for if

$$x - \chi(\xi)$$

we may substitute for x and obtain the distribution of ξ in the form

$$df = \phi\{\chi(\xi)\} \frac{d\chi}{d\xi}\, d\xi.$$

Obviously the form of the distribution has changed; thus, if we know the frequency distribution of the time in which a number of men run 100 yards, we may derive the distribution of their velocities, which will be a different distribution, obtained simply by transforming df as a differential element. In particular we must notice that the mean of the distribution is not invariant for such transformations, thus, if $\bar{x}$ and $\bar{\xi}$ are the means of their respective distributions, we shall not in general find that

$$\bar{x} = \chi(\bar{\xi}).$$

Similarly, the *mode*, that is, the point, if there is one, at which ϕ has a maximum for variation of x, will not be invariant, for the equations

$$\frac{d^2f}{dx^2} = 0, \quad \frac{d^2f}{d\xi^2} = 0$$

will not normally be satisfied by corresponding values. The central measure which is invariant, at least if $d\chi/d\xi$ is positive for all values, is the *median*, the value which divides the total frequency into two equal halves. For this point $f = \frac{1}{2}$, and the values of x and ξ will be necessarily in agreement. The same will be true of all other points defined by the value of f, so that we may have deciles, centiles, etc., dividing the frequency into 10 or 100 equal parts, and these will be invariant for any transformation for which $d\chi/d\xi$ is always positive.

All the above applies with no essential change to the more general case in which we have several observable variates $x, y, z, \ldots$ in place of one.

The general statement of the inverse type of argument is as follows; we shall first cloak its fallacy under an hypothesis, and then examine it as an undisguised assumption.

Suppose that we know that the population from which our observations were drawn had itself been drawn at random from a super-population of known specification; that is, suppose that we have *a priori* knowledge that the probability that $\theta_1, \theta_2, \theta_3, \ldots$ shall lie in any defined infinitesimal range $d\theta_1 d\theta_2 d\theta_3 \ldots$ is given by

$$dF = \Psi(\theta_1, \theta_2, \theta_3, \ldots) d\theta_1 d\theta_2 d\theta_3 \ldots,$$

then the probability of the successive events (*a*) drawing from the super-population a population with parameters having the particular values $\theta_1, \theta_2, \theta_3, \ldots$ and (*b*) drawing from such a population the sample values $x_1, \ldots, x_n$, will have a joint probability

$$\Psi(\theta_1, \theta_2, \theta_3, \ldots) d\theta_1 d\theta_2 d\theta_3 \ldots \times \prod_{p=1}^{n} \{\phi(x_p, \theta_1, \theta_2, \theta_3, \ldots) dx_p\}.$$

If we integrate this over all possible values of $\theta_1, \theta_2, \theta_3, \ldots$ and divide the original expression by the integral we shall then have a perfectly definite value for the probability (in view of the observed sample and of our *a priori* knowledge) that $\theta_1, \theta_2, \theta_3, \ldots$ shall lie in any assigned limits.

This is not inverse probability strictly speaking, but a perfectly direct argument, which gives us the frequency distribution of the population parameters θ, from which we may, if we like, calculate their means, modes, medians or whatever else might be of use.

The peculiar feature of the inverse argument proper is to say something equivalent to "We do not know the function Ψ specifying the super-population, but in view of our ignorance of the actual values of θ we may take Ψ to be constant." Perhaps we might add that all values of θ being equally possible their probabilities are by definition equal; but however we might disguise it, the choice of this particular *a priori* distribution for the θ's is just as arbitrary as any other could be. If we were, for example, to replace our θ's by an equal number of functions of them, θ_1', θ_2', θ_3', ... all objective statements could be translated from the one notation to the other, but the simple assumption $\Psi(\theta_1, \theta_2, \theta_3, ...) = \text{constant}$ may translate into a most complicated frequency function for

$$\theta_1', \theta_2', \theta_3', ...$$

If, then, we follow writers like Boole, Venn, and Chrystal in rejecting the inverse argument as devoid of foundation and incapable even of consistent application, how are we to avoid the staggering falsity of saying that however extensive our knowledge of the values of x may be, yet we know nothing and can know nothing about the values of θ? Inverse probability has, I believe, survived so long in spite of its unsatisfactory basis, because its critics have until recent times put forward nothing to replace it as a rational theory of learning by experience.

The first point to be made belongs to the theory of statistical estimation; it has nothing to do with inverse probability, save for the historical accident that it was developed by Gauss in terms of that theory.

If we make the assumption that $\Psi(\theta_1, \theta_2, \theta_3, ...) = \text{constant}$, and if then we ignore everything about the inverse probability distribution so obtained except its mode or point at which the ordinate is greatest, we have to maximise

$$\prod_{p=1}^{n} \{\phi(x_p, \theta_1, \theta_2, \theta_3, ...)\}$$

for variations of θ_1, θ_2, θ_3, ...; and the result of *this* process will be the same whether we use the parameters θ_1, θ_2, θ_3, ... or any functions of them, θ_1', θ_2', θ_3', ... Two wholly arbitrary elements in this process have in fact cancelled each other out, the non-invariant process of taking the mode, and the arbitrary assumption that Ψ is constant. The choice of the mode is thinly disguised as that of "the most probable value," whereas had the inverse probability distribution any objective reality at all we should certainly, at least for a single parameter, have preferred to take the mean or the median value. In fact neither of these two processes has a logical justification, but each is necessary to eliminate the errors introduced by the other.

The process of maximising $\Pi(\phi)$ or $S(\log\phi)$ is a method of estimation known as the "method of maximum likelihood"; it has in fact no logical connection with inverse probability at all. The facts that it has been accidentally associated with inverse probability, and that when it is examined objectively in respect of the properties in random sampling of the estimates to which it gives rise, it has shown itself to be of supreme value, are perhaps the sole remaining reasons why that theory is still treated with respect. The function of the θ's maximised is not however a probability and does not obey the laws of probability; it involves no differential element $d\theta_1 d\theta_2 d\theta_3 \ldots$; it does none the less afford a rational basis for preferring some values of θ, or combination of values of the θ's, to others. It is, just as much as a probability, a numerical measure of rational belief, and for that reason is called the *likelihood* of $\theta_1, \theta_2, \theta_3, \ldots$ having given values, to distinguish it from the probability that $\theta_1, \theta_2, \theta_3, \ldots$ lie within assigned limits, since in common speech both terms are loosely used to cover both types of logical situation.

If A and B are mutually exclusive possibilities the probability of "A or B" is the sum of the probabilities of A and of B, but the likelihood of A or B means no more than "the stature of Jackson or Johnson"; you do not know what it is until you know which is meant. I stress this because in spite of the emphasis that I have always laid upon the difference between probability and likelihood there is still a tendency to treat likelihood as though it were a sort of probability.

The first result is thus that there are two different measures of rational belief appropriate to different cases. Knowing the population we can express our incomplete knowledge of, or expectation of, the sample in terms of probability; knowing the sample we can express our incomplete knowledge of the population in terms of likelihood. We can state the relative likelihood that an unknown correlation is $+0\cdot6$, but not the probability that it lies in the range $\cdot595$—$\cdot605$.

There are, however, certain cases in which statements in terms of probability can be made with respect to the parameters of the population. One illustration may be given before considering in what ways its logical content differs from the corresponding statement of a probability inferred from known *a priori* probabilities. In many cases the random sampling distribution of a statistic, T, calculable directly from the observations, is expressible solely in terms of a single parameter, of which T is the estimate found by the method of maximum likelihood. If T is a statistic of continuous variation, and P the probability that T should be less than any specified value, we have then a relation of the form

$$P = F(T, \theta).$$

If now we give to P any particular value such as ·95, we have a relationship between the statistic T and the parameter θ, such that T is the 95 per cent. value corresponding to a given θ, and this relationship implies the perfectly objective fact that in 5 per cent. of samples T will exceed the 95 per cent. value corresponding to the actual value of θ in the population from which it is drawn. To any value of T there will moreover be usually a particular value of θ to which it bears this relationship; we may call this the "fiducial 5 per cent. value of θ" corresponding to a given T. If, as usually if not always happens, T increases with θ·for all possible values, we may express the relationship by saying that the true value of θ will be less than the fiducial 5 per cent. value corresponding to the observed value of T in exactly 5 trials in 100. By constructing a table of corresponding values, we may know as soon as T is calculated what is the fiducial 5 per cent. value of θ, and that the true value of θ will be less than this value in just 5 per cent. of trials. This then is a definite probability statement about the unknown parameter θ, which is true irrespective of any assumption as to its *a priori* distribution.

Fiducial 5 °/₀ ρ	95 °/₀ r	Fiducial 5 °/₀ ρ	95 °/₀ r	Fiducial 5 °/₀ ρ	95 °/₀ r
− ·995055	− ·968551	− ·761594	+ ·145340	+ ·761594	+ ·989816
− ·993963	− ·961623	− ·716298	+ ·270475	+ ·800499	+ ·991770
− ·992632	− ·953179	− ·664037	+ ·388574	+ ·833655	+ ·993335
− ·991007	− ·942894	− ·604368	+ ·496089	+ ·861723	+ ·994593
− ·989027	− ·930375	− ·537050	+ ·590725	+ ·885352	+ ·995608
− ·986614	− ·915151	− ·462117	+ ·671557	+ ·905148	+ ·996427
− ·983675	− ·896661	− ·379949	+ ·738849	+ ·921669	+ ·997091
− ·980096	− ·874240	− ·291313	+ ·793711	+ ·935409	+ ·997628
− ·975743	− ·847110	− ·197375	+ ·837715	+ ·946806	+ ·998066
− ·970452	− ·814372	− ·099668	+ ·872590	+ ·956237	+ ·998421
− ·964028	− ·775019	0	+ ·900000	+ ·964028	+ ·998711
− ·956237	− ·727916	+ ·099668	+ ·921432	+ ·970452	+ ·998646
− ·946806	− ·671918	+ ·197375	+ ·938146	+ ·975743	+ ·999139
− ·935409	− ·605881	+ ·291313	+ ·951174	+ ·980096	+ ·999296
− ·921669	− ·528824	+ ·379949	+ ·961338	+ ·983675	+ ·999424
− ·905148	− ·440127	+ ·462117	+ ·969286	+ ·986614	+ ·999529
− ·885352	− ·339761	+ ·537050	+ ·975519	+ ·989027	+ ·999615
− ·861723	− ·228562	+ ·604368	+ ·980424	+ ·991007	+ ·999685
− ·833655	− ·108446	+ ·664037	+ ·984298	+ ·992632	+ ·999742
− ·800499	+ ·017528	+ ·716298	+ ·987371	+ ·993963	+ ·999789
− ·761594	+ ·145340	+ ·761594	+ ·989816	+ ·995055	+ ·999827

For example, if r is a correlation derived from only four pairs of observations, and ρ is the correlation in the population from which the sample was drawn, the relation between ρ and the 95 per cent. value of r is given in the following table, which has been calculated, from the distribution formula I gave in 1915, by Miss F. E. Allan. From the table we can read off the 95 per cent. r for any given ρ, or equally the fiducial 5 per cent. ρ for any given r. Thus if a value $r = \cdot99$ were obtained from the sample, we should have a fiducial 5 per cent. ρ equal to about $\cdot765$. The value of ρ can then only be less than $\cdot765$ in the event that r has exceeded its 95 per cent. point, an event which is known to occur just once in 20 trials. In this sense ρ has a probability of just 1 in 20 of being less than $\cdot765$. In the same way, of course, any other percentile in the fiducial distribution of ρ could be found or, generally, the fiducial distribution of a parameter θ for a given statistic T may be expressed as

$$df = -\frac{\partial}{\partial\theta} F(T, \theta)\, d\theta,$$

while the distribution of the statistic for a given value of the parameter is

$$df = \frac{\partial}{\partial T} F(T, \theta)\, dT.$$

I imagine that this type of argument, which supplies definite information as to the probability of causes, has been overlooked by the earlier writers on probability, because it is only applicable to statistics of continuous distribution, and not to the cases in regard to which the abstract arguments of probability theory were generally developed, in which the objects of observation were classified and counted rather than measured, and in which therefore all statistics have discontinuous distributions. Now that a number of problems of distribution have been solved, for statistics having continuous distribution, arguments of this type force themselves on our attention; and I have recently received from the American statistician, Dr M. Ezekiel, graphs giving to a good approximation the fiducial 5 per cent. points of simple and multiple correlations for a wide range of cases. It is therefore important to realise exactly what such a probability statement, bearing a strong superficial resemblance to an inverse probability statement, really means. The fiducial frequency distribution will in general be different numerically from the inverse probability distribution obtained from any particular hypothesis as to *a priori* probability. Since such an hypothesis may be true, it is obvious that the two distributions must differ not only numerically, but in their logical meaning. It would be perfectly possible, for example, to find an *a priori* frequency distribution for ρ such that the inverse probability that ρ is less than ·765 when $r = $ ·99 is not 5 but 10 in 100. In concrete terms of frequency this would mean that if we repeatedly selected a population at random, and from each population selected a sample of four pairs of observations, and rejected all cases in which the correlation as estimated from the sample (r) was not exactly ·99, then of the remaining cases 10 per cent. would have values of ρ less than ·765. Whereas apart from any sampling for ρ, we know that if we take a number of samples of 4, from the same or from different populations, and for each calculate the fiducial 5 per cent. value for ρ, then in 5 per cent. of cases the true value of ρ will be less than the value we have found. There is thus no contradiction between the two statements. The fiducial probability is more general and, I think, more useful in practice, for in practice our samples will all give different values, and therefore both different fiducial distributions and different inverse probability distributions. Whereas, however, the fiducial values are expected to be different in every case, and our probability statements are relative to such variability, the inverse probability statement is absolute in form and really means something different for each different sample, unless the observed statistic actually happens to be exactly the same.

Appendix A
English Translations of Papers and Book Extracts of Historical Interest

Presented below is a list of the relatively few English translations we have been able to find of articles or book extracts that are of interest in the history of statistics. Nearly all the translations are accompanied by commentary, often by the translator or the editor. Page numbers of the translations do not include commentary. Wherever possible we name the translator and also the commentator if other than translator or editor. In length, the translations vary from Legendre's one-page 1820 attack on Gauss to the 124 pages of Laplace's *Essai*, extensively annotated by A. I. Dale. Dates range from Galileo, around 1620, to de Finetti in 1976.

Part of our list is drawn from the references in Hald (1990, 1998). In some cases Hald provides additional details on translations into other languages. His references also include excellent information on translations of books of historical importance, so that we have not attempted to list book translations.

Special mention must be made of recent (1998–2000) translations from the Russian by O. Sheynin. Five collections of papers on probability and statistics have been published on microfiche by Haensel-Hohenhausen, Egelsbach, in their Deutsche Hochschulschriften series. Specific titles and ISBN numbers are: *From Markov to Kolmogorov* (3 8267–2514), *From Davidov to Romanovsky* (3–8267–2579), Sheynin, *Russian Papers on History of Probability and Statistics* (3–8267–2621), *From Bortkiewicz to Kolmogorov* (3–8267–2656), Chebyshev, *Theory of Probability, Lectures of 1880* (3–8267–2665), and soon to be published, *From Dan. Bernoulli to Steklov*. Other authors featured include: Bernstein, Chuprov, Gnedenko, Idelson, Khinchin, Liapunov, Markov, Nekrasov, Ondar, Paevsky, Sarymsakov, Slutsky, Smirnov, and Urlanis.

Also given below is the short list of collected papers and source books cited in the above references. However, these publications include primarily papers originally published in English, together with commentary. Pearson and Kendall (1970) and Kendall and Plackett (1977) consist mainly of review and discussion articles on many aspects of the history of statistics. Somewhat of an outlier is Kolmogorov (1992), which is a translation from a Russian edition of Kolmogorov's papers on probability and mathematical statistics. His papers were first published in French, German, Italian, or Russian.

Papers and Book Extracts

Bernoulli, D. (1738). Specimen theoriae novae de mensura sortis. *Comment. Acad. Sci. Imp. Petrop.*, **5**, 175–192 (1730–1731). *Werke*, **2**, 223–234. Translated by L. Sommer (1954) as Exposition of a new theory on the measurement of risk. *Econometrica*, **22**, 23–35.

Bernoulli, D. (1766). Essai d'une nouvelle analyse de la mortalité causée par la petite vérole.... *Hist. Mém. Acad. R. Sci. Paris*, 1–45 (1760). *Werke*, **2**, 235–267. Translated by L. Bradley (1971) as An attempt at a new analysis of the mortality caused by smallpox and of the advantages of inoculation to prevent it. In *Smallpox Inoculation: An Eighteenth Century Mathematical Controversy*. Adult Education Dept., University of Nottingham.

Bernoulli, D. (1778). Diiudicatio maxime probabilis plurium observationem discrepantium.... *Acta Acad. Sci. Imp. Petrop.*, **1**, 3–23 (1777). *Werke*, **2**, 361–375. Translated by C.G. Allen (1961) as The most probable choice between several discrepant observations and the formation therefrom of the most likely induction. *Biometrika*, **48**, 3–18. Commentary by M.G. Kendall. Reprinted in Pearson and Kendall (1970, pp. 157–172). See also Euler (1778). A 1769 manuscript with the same title but substantial differences has been translated by S. Lerer, M. McGrade, and S. Stigler in D. Pollard, E. Torgerson, and G. Yang (eds.) (1997). *Festschrift for Lucien Le Cam*, pp. 359–367. Commentary by S.M. Stigler.

Bernoulli, James. (1713). *Ars conjectandi*. Thurnisius, Basilea. Chapter V. Part 2. On the theory of combinations. Translated by M.M. Taylor. Smith (1929, pp. 272–277). All of Part 2 translated by Maseres (1795). See also Note [1], p. 135 of Edwards (1987).

Borel, E. (1924). A propos d'un traité de probabilités. In *Valeur pratique et philosophie des probabilités*, pp. 134–146. Translated by H.E. Smokler, as A propos of a treatise on probability [by Keynes]. Kyburg and Smokler (1964, pp.47–60).

Cardano, G. (c. 1564) *Liber de ludo aleae*. First printed in *Opera Omnia*, Vol. 1, 1663. Translated by S.H. Gould as *The Book on Games of Chance*. Ore (1953, pp. 183–241).

Chebyshev, P.L. (1867). Des valeurs moyennes. *Liouville's J. Math. Pures Appl.*, [2] **12**, 177–184, Translated by H. Walker as On the mean values. Smith (1929, pp. 581–587).

Craig, J. (1699). *Theologiae Christianae Principia Mathematica*. Reprinted in part and translated as Craig's rules of historical evidence. *History and Theory: Studies in the Philosophy of History*, Suppl. **4**, 1–31 (1964).

Euler, L. (1778). Observationes in praecedentem dissertationem illustris Bernouli. *Acta Acad. Sci. Imp. Petrop.*, **1**, 24–33 (1777). *Opera Omnia*, [1] **7**, 280–290. Translated by C.G. Allen (1961) as Observations on the foregoing dissertation of Bernoulli. *Biometrika*, **48**, 13–18. See also Bernoulli (1778).

Fermat, P. (1654). Correspondence with Pascal. See Pascal (1654).

Finetti, B. de (1937). La prévision.... *Ann. Inst. Poincaré*, **7**, 1–68. Translated by H.E. Kyburg, Jr., as Foresight: Its logical laws, its subjective sources. Kyburg and Smokler (1980, pp. 57–118). First four (of six) chapters reproduced in Kotz and Johnson (1992a, pp. 135–174) with commentary by R.E. Barlow.

Finetti, B. de (1976). *Scientia*, **111**, 283–303. Translated by I. McGilvray as Probability: Beware of falsifications! Kyburg and Smokler (1980, pp. 195–224).

Galileo, G. (c. 1620). *Sopra la scoperte dei dadi (On a Discovery Concerning Dice)*. Translated by E.H. Thorne in F.N. David (1962, pp. 192–195).

Gauss, C.F. (1957). *Gauss's Work (1809–1826) on the Theory of Least Squares*. Translated by H.F. Trotter from the 1855 French translation by J. Bertrand of Gauss's Latin and German. Tech. Rep. No. 5. Statist. Techniques Res. Gp., Princeton University.

Gergonne, J.D. (1815). Application de la méthode des moindres quarrés à l'interpolation des suites. *Ann. Math. Pures et appl.*, **6**, 242–252. Translated by R. St. John and S.M. Stigler (1974) as The application of the method of least squares to the interpolation of sequences. *Historia Mathematica*, **1**, 439–447. Commentary by S.M. Stigler.

Gnedenko, B.V. (1943). Sur la distribution limite du terme maximum d'une série aléatoire. *Ann. Math.*, **44**, 423–453. Translated by N.L. Johnson as On the limiting distribution of the maximum term in a random series. Kotz and Johnson (1992a, pp. 195–225). Commentary by R.L. Smith.

Kolmogorov, A.N. (1933). Sulla determinazione empirica di una legge di distribuzione. *Ist. Ital. Attuari*, *G.*, **4**, 1–11. Translated by Q. Meneghini as On the empirical determination of a distribution function. Kotz and Johnson (1992b, pp. 106–113). Commentary by M.A. Stephens.

Laplace, P.S. (1774). Mémoire sur la probabilité des causes par les événements. *Mém. Acad. R. Sci.*, **6**, 621–656. *Oeuvres complètes*, **8**, 27–65. Translated, with commentary, by S. Stigler (1986) as Memoir on the probability of the causes of events. *Statist. Sci.*, **1**, 364–378.

Laplace, P.S. (1812). *Théorie analytique des probabilités*. Courcier, Paris. *Oeuvres complètes*, **7**. Ch. 4, Sections 18–20 translated by J. Gÿs as On the probability of the errors in the mean results of a great number of observations and on the most advantageous mean result. Smith (1929, pp. 588–604).

Laplace, P.S. (1825). *Essai philosophique sur les probabilités*, 5th edn., reprinted with notes by B. Bru (1986). Bourgois, Paris. Translated, with commentary, by A.I. Dale. (1995) as *Philosophical Essay on Probabilities*. Springer, New York.

Legendre, A.M. (1805). *Nouvelles méthodes pour la détermination des orbites des comètes*. Courcier, Paris. Initial four pages of the appendix on

206

the method of least squares translated by H.A. Ruger and H.M. Walker in D.E. Smith (1929, pp. 576–579).

Legendre, A.M. (1820). Note par M..., pp. 79–80 of Second Supplement of 2nd edn. of *Nouvelles méthodes....* Translated by S.M. Stigler (1977, pp. 33–34), with commentary, in An attack on Gauss published by Legendre in 1820. *Historia Mathematica,* **4,** 31–35.

Maseres, F. (1795). *Mr. James Bernoulli's Doctrine of Permutations and Combinations, and Some Other Useful Mathematical Tracts.* White, London.

Moivre, A. de (1712). De mensura sortis.... *Phil. Trans. R. Soc. London,* **27,** 213–264. Translated by B. McClintock (1984) as On the measurement of chance, or, on the probability of events in games depending upon fortuitous chance. *Intern. Statist. Rev.,* **52,** 237–262. Commentary by A. Hald.

Moivre, A. de (1733). Approximatio ad summam terminorum binomii $\overline{a+b}\,^n$ in seriem expansi. Circulated privately. Translated by the author as A method of approximating the sum of the terms of the binomial $\overline{a+b}\,^n$ expanded into a series from whence are deduced some practical rules to estimate the degree of assent which is to be given to experiments. *The Doctrine of Chances,* 2nd edn. (1738, pp. 235–243), 3rd edn. (1756, pp. 243–254). Reproduced in Smith (1929, pp. 566–575). Commentary by R.H. Daw and E.S. Pearson in Kendall and Plackett (1977, pp. 63–66).

Pascal, B. (1654). Correspondence with Fermat. Reprinted in *Oeuvres complètes,* 1779 and later editions. Translated by V. Sanford in Smith (1929, pp. 546–565) and by M. Merrington in F.N. David (1962, pp. 229–253). But see Note [7], p. 68 and Appendix 1 of Edwards (1987) for an important correction.

Poisson, S.D. (1837). *Recherches sur la probabilité des jugements en matière criminelle et en matière civile précédées des règles génerales du calcul des probabilités.* Pp. 189–190 and 205–207 translated, with commentary, by S.M. Stigler (1982) in Poisson on the Poisson distribution. *Statist. Probab. Lett.,* **1,** 33–35.

Steffensen, J.F. (1930). *Mat. Tidskr.* B, 19–23. Translated from the Danish by P. Guttorp as On the probability that the offspring dies out. *Intern. Statist. Rev.,* **63,** 239–242. Commentary by K. Albertsen.

Thiele, T.N. (1871). On a mathematical formula to express the rate of mortality throughout the whole life, tested by a series of observations made use of by the Danish Life Insurance Company of 1871. Translated from the Danish by T.B. Sprague (1871). *J. Inst. Actuaries,* **16,** 313–329.

Collections and Source Books

David, F.N. (1962). *Games, Gods and Gambling*. Hafner, New York.

Kendall, M.G. and Plackett, R.L. (eds.) (1977). *Studies in the History of Statistics and Probability*, Vol. II. Griffin, London.

Kolmogorov, A.N. (1992). *Selected Works of A.N. Kolmogorov*, Vol. II. *Probability Theory and Mathematical Statistics*. Edited by A.N. Shiryayev and translated from the Russian by G. Lindquist. Kluwer, Dordrecht, Netherlands.

Kotz, S. and Johnson, N.L. (eds.) (1992a, b; 1997). *Breakthroughs in Statistics*, Vols. I-III. Springer, New York.

Kyburg, H.E., Jr. and Smokler, H.E. (eds.) (1964). *Studies in Subjective Probability*. Wiley, New York.

Kyburg, H.E., Jr. and Smokler, H.E. (eds.) (1980). *Studies in Subjective Probability*, 2nd edn. Krieger, Huntington, NY.

Newman, J.R. (ed.) (1956). *The World of Mathematics*, Four volumes. Simon and Schuster, New York.

Ore, O. (1953). *Cardano, the Gambling Scholar*. Princeton University Press.

Pearson, E.S. and Kendall, M.G. (eds.) (1970). *Studies in the History of Statistics and Probability*. Hafner, Darien, CT.

Smith, D.E. (1929). *A Source Book in Mathematics*. McGraw-Hill, New York.

Other References

Edwards, A.W.F. (1987). *Pascal's Arithmetical Triangle*. Griffin, London; Oxford University Press, New York.

Hald, A. (1990). *A History of Probability and Statistics and Their Applications before 1750*. Wiley, New York.

Hald, A. (1998). *A History of Mathematical Statistics from 1750 to 1930*. Wiley, New York.

Appendix B

First (?) Occurrence of Common Terms in Statistics and Probability

H. A. David

1 Introduction

In this appendix an attempt is made to list the first occurrence in print of terms commonly used in statistics or probability. The list is motivated by the recognition that the coining of a term is often an important step in fixing and propagating a concept. Existence of a term simplifies indexing and information retrieval. Also, it seems desirable to be aware of the origin of terms, especially those used in basic courses. But it should be understood that the date of the term does not necessarily signify the beginning of the underlying idea.

Of course, establishing a first occurrence is hazardous; hence the question mark in the title of this appendix. Some listings are, in fact, quite unlikely to be firsts, but the dates given should provide good upper bounds. The selection of terms is necessarily somewhat personal, and it will be easy for the reader to spot absences. Terms honoring an individual, such as "Poisson distribution," are included only when very familiar. Purely mathematical terms are excluded.

When a concept is much older than its present name, there may be an interesting word evolution to pursue. For example, "frequency curve" and "frequency function," which long held sway, eventually were shouldered aside by "probability density" and "probability density function." Terms judged to be obsolete or obsolescent are not included in the list, nor are a few basic old terms whose current technical use is essentially a refinement of their everyday meaning, e.g., error, expectation, hypothesis, independence, mean, outlier, probability, random, sample, and stochastic. For these the reader may wish to refer to the 20-volume *Oxford English Dictionary* (OED) published in 1989. However, many technical terms involving the above words are listed, e.g., error of the first kind, null hypothesis, prior probability, random variable, and stochastic process.

The focus is on terms in English, but when the English term is known to me to have been preceded by a foreign term, the earliest (?) foreign term is listed in parentheses, immediately following the English term, e.g., "random variable" is followed by "(variabile casuale)." It is worth noting that in this case the English term is not an obvious translation and "chance variable," "stochastic variable," and "aleatory variable" have all been used. When the English term is just a trivial translation, less effort has been made to trace its first occurrence.

2   Methods Used

A good starting point in the search for a first occurrence is the big OED, already mentioned. Together with its three small Addition Volumes it accounts for an estimated 20% of our list, mainly single-word terms. The OED is impressive, but not infallible in giving the earliest citation. The *Encyclopedia of Statistical Sciences* (Kotz, Johnson, and Read, 1982–1989) sometimes gives the first occurrence and, much more often, provides a lead. Also occasionally useful are Kruskal and Tanur (1978), Marriott (1990), and the subject indexes to *Biometrika*, 1901–1953, to the *Annals of Mathematical Statistics*, 1930–1960, and to *Biometrics*, 1945–1964. For more recent terms, permuted title indexes are useful aids: *JASA*, 1955–1991, *IMS Scientific Journals*, 1960–1989, and, of course, the *Current Index to Statistics*, 1975–. Some books stand out for present purposes: Yule (1911), Pearson's (1914–1930) *Life of Francis Galton*, Wilks (1943), Cramér (1946), Fisher (1971), Harter (1978), and Lehmann (1983).

Special mention must be made of Helen Walker's *Studies in the History of Statistical Method* (Walker, 1929). This contains a chapter, entitled "The Origin of Certain Technical Terms Used in Statistics," having the same aim as this appendix, except for an emphasis on educational statistics. Her list of 120 terms, while overlapping only modestly with this one of c. 450, has enabled me to make a number of improvements. I am indebted to Erich Lehmann for drawing my attention to Walker's list upon seeing some early work of mine.

With no better starting point for a particular term, one will want to turn to subject indexes. These may list the term, but the articles to which one is referred may deal with the relevant subject matter without actually using the term. If the term does occur, one must check likely references in the articles for yet earlier occurrences. Continuing this process evidently ensures only something like a local minimum. Yet it often becomes apparent from statements made that one probably has reached the earliest date. A further check is provided by perusing relevant books published prior to the presumed first occurrence date.

That there can be no guarantee of success is well illustrated by "normal" as applied to the distribution. The OED leads one to Pearson (1893), a citation bolstered by Pearson (1920) where we read: "Many years ago I called the Laplace–Gaussian curve the *normal* curve, which name, while it avoids an international question of priority, has the disadvantage of leading people to believe that all other distributions of frequency curves are in one sense or another abnormal." Actually, Pearson need not have issued this uncharacteristic apology, since the term had already been used by his hero, Francis Galton, and others. Kruskal and Stigler (1997) have given a wide-ranging, interesting discussion of the word "normal," crediting C. S. Peirce (1873, p. 206) with the first use of the term, to describe what we now call the "standard normal."

In nearly every case the cited reference itself (or a reprinting) has been examined to ensure that it indeed includes the term of interest and not just the concept. One needs to check as well that the term has essentially its current meaning. For example, for the term "bootstrap" we give Efron (1979). However, Huang (1972) writes of a "bootstrap procedure" in reference to an article by Robbins (1951) on a compound decision procedure that contains the phrase, "lift ourselves by our bootstraps." Efron and Tibshirani (1993) state that the saying "to pull oneself up by one's bootstraps" is widely thought to have originated in the 18th century *Adventures of Baron Munchausen*, written in English by Rudolf Erich Raspe. The Baron had fallen to the bottom of a deep lake. Just when all seemed lost, he thought to pick himself up by his own bootstraps.

Different problems are illustrated by the term "P value" or "P-value." This is "the smallest level at which the observations are significant in a particular direction" (Gibbons and Pratt, 1975). The idea goes back at least to Laplace (Stigler, 1986, p. 151), but numerous alternative terms have been used and to some extent are still being used: probability level, sample level of significance, observed significance level, significance probability, descriptive level of significance, critical level, significance level, prob-value, and associated probability. Pearson (1900) already used P in this kind of context. Berkson (1942) writes of "the P of the test." The first use of P value I have spotted is in Deming (1943), a reference supplied by N. J. Cox. This is the only entry on this subject included in our list. Here and in some other instances a decision had to be made as to when a term could be regarded as having reached essentially its present form.

In most cases, when competing terms have been around for a long time, one of them establishes its dominance. A striking exception, to this day, is in the term for $f(x) = P(X = x)$ when X is a discrete rv. Wilks (1943), who first (?) called $f(x)$ the probability density function when X is a continuous rv, avoids any term in the discrete case, as does, e.g., Cramér (1937, 1946). Other authors, however, use probability density function for both continuous and discrete rvs, e.g., Hogg and Craig (1959, p. 13). Lehmann (1983, p. 16) speaks of a "probability density with respect to counting measure." Two often used terms, included in our list, are "probability function" and "probability mass function" that go back at least to Aitken (1939) and Parzen (1960). Even "frequency function" survives in Stuart and Ord (1987, p. 13) to cover both discrete and continuous rvs.

3   Further Comments on Selected Terms

We begin with "statistics" itself, not a very old term, even in its early political connotation due to German writers. The OED has the citation

> **1770** W. Hooper tr. Bielfeld's Elem. Universal Educ. III xiii 269. The science, that is called statistics, teaches us what is the political arrangement of all the modern states of the known world.

An interesting treatment of how the term evolved is presented in Yule's (1911) admirable text. See also Westergaard (1932).

Apart from "statistics," the oldest term in our list is "Latin square," introduced as "quarré latin" in 1782 by Euler, in the French spelling of the time. Cayley (1890) may have been the first to use the English version. "Graeco-Latin square" is due to Fisher and Yates (1934), but the use of Latin and Greek letters goes back to Euler. Next oldest is "method of least squares," a term coined by Legendre in 1805 as "méthode des moindres quarrés." The method was known to Gauss a little earlier but he did not publish it then. There is no doubt that Legendre deserves full credit for the name, which caught on quickly. Actually, "minimum" and "small" were the early English translations of "moindres" until Ivory (1825), who still referred to "the method of the least squares," a literal translation. See Stigler (1986, p. 15) and also Merriman (1877–1882) and Plackett (1972). The two newest terms are "Gibbs Sampler" (Geman and Geman, 1984) and "computer-intensive" (Diaconis and Efron, 1983).

Brownian motion. Under *Brownian* the OED gives only the now superseded *Brownian movement* which it traces back to 1872. Curiously, Brown in 1828 used *motion* repeatedly in a passage cited by Brush (1968). A. M. Hughes, Chief Science Editor of the OED, informs me that *Brownian motion* occurs in the OED entry for *pedetic* and was used by Ramsay (1892). *Brownian* will be extended to *Brownian motion* in the revised version of the OED now in preparation.

Cumulant. This term, replacing *semi-invariant*, was first published by Fisher and Wishart (1931). In a private communication S. Stigler has pointed out that the term was actually suggested to Fisher by Hotelling. See the latter's review of the 4th edition (1932) of *Fisher's Statistical Methods for Research Workers* in *J. Amer. Statist. Ass.*, **28**, 374–375 (1933).

Exchangeable. The concept of exchangeability has been most thoroughly explored by de Finetti. However, J. Haag and W.E. Johnson anticipated de Finetti in some respects (around 1924). The term itself was introduced by Fréchet in 1939. See Barlow's (1991) introduction to an English translation of Finetti (1937) referenced in Appendix A.

Law of the iterated logarithm. Kolmogoroff (1929) states that the term was first used in Khintchine's 1927 Russian text *Principles of Prob-*

ability Theory. Uspensky (1937, p. 204) writes of "the law of the repeated logarithm."

Monte Carlo method. According to a report in the journal *Mathematical Tables and Other Aids to Computation* (1949, p. 546) both method and name were apparently first suggested by John von Neumann and S. M. Ulam. H. Riedwyl has informed me of the Metropolis and Ulam reference listed here.

Nearest neighbor. I could trace no earlier use in the regular statistical literature than Halperin (1960, p. 1064), where the listed reference to Chandrasekar (1943) is given. In statistics the idea goes back to a 1937 paper by Papadakis. This was elaborated by Bartlett in 1938, who later helped to popularize the term. See the entry "nearest-neighbor methods" in the *Encyclopedia of Statistical Sciences* (Kotz et al. 1982–1989).

Petersburg paradox. Jorland (1987) traces the evolution of the paradox, pointing out that d'Alembert "christened" it in 1768. (*Opuscules mathématiques*, Vol. IV, p. 78). There one finds "le problème de Petersbourg." Keynes (1921), cited in our list, seems unlikely to have been the first to introduce "paradox." Fry (1928) gives an excellent brief discussion of the *St. Petersburg paradox.*

Sign test. As is well known, the sign test may be regarded as the first test of significance, going back to Arbuthnott (1712), the second paper in this volume. A. Hald has recently pointed out to me that the test is discussed by Helmert (1905), who actually uses the term, in German. The English term seems to have developed independently (Stewart, 1941).

Signed-rank test. This ingenious test is due to Wilcoxon (1945). The clever and helpful term was coined by Tukey (1949) in an unpublished, but repeatedly cited technical report (e.g., Lehmann, 1953, p. 43).

Weighted least squares. Inclusion of this term was suggested by A. Hald. Although the application of least squares to observations with unequal variances goes back at least to Gauss, I was unable to find the term any earlier than in K. Pearson (1920).

For comments on *Unbiased* to *Uniformly minimum variance unbiased* see David (1995, p. 122) and for *Game theory* see David (1998, p. 36).

4   The List of Citations

With rare exceptions a term is given in only one grammatical form, most often as a noun, even if first occurring in a different form. Sometimes the listed term differs in unimportant ways from its form in the reference cited. This is indicated by "(essentially)" following the term. For example, "percentage point" actually is "per cent point" in Fisher (1925a).

American spelling is used. A special case is that of homo- and heterosc(sk)edasticity, all four of which are included in our list. McCulloch

(1985) makes a strong case for k, based on its Greek kappa root. His argument is even more compelling for American English than for British English, which also has "sceptic." Nevertheless, many American authors have followed Karl Pearson's (1905) spelling. McCulloch gives Valavanis (1959) as the earliest user of k. It is surprising that the editor of *Biometrika* chose c over k.

To assist the reader who wishes to follow up on any of the citations, a page number is given in nearly every case. A large proportion of the citations is due to leaders in our field, whose collected works, usually also included in the list of references, may sometimes be more readily available. The page number in the reprinting may then have to be inferred.

α-design	Patterson, H.D. and Williams, E.R. (1976, p. 83)
Additivity (in ANOVA)	Eisenhart, C. (1947, p. 10)
Admissibility	Wald, A. (1939, p. 301)
Alias	Finney, D.J. (1945, p. 292)
Analysis of variance	Fisher, R.A. (1918b, p. 219)
—— of covariance	Bailey, A.L. (1931, title)
Ancillary statistics	Fisher, R.A. (1925b, p. 724)
Angular transformation	Fisher, R.A. and Yates, F. (1938, p. vii)
Association	Yule, G. U. (1900, title)
—— scheme	Bose, R. C. and Shimamoto, T. (1952, p. 154)
Asymptotic distribution	Wald, A. and Wolfowitz, J. (1940, p. 151)
—— efficiency	Wald, A. (1948, p. 41)
—— normality	Neumann, J. von (1941, p. 379)
—— relative efficiency (of tests)	Noether, G.E. (1950, p. 241)[1]
—— variance-covariance matrix	Wilks, S.S. (1943, p. 142)
Asymptotically most powerful test	Wald, A. (1941, title)
Autocorrelation	Wold, H. (1938, p. 6)
Autoregression	Wold, H. (1938, p. 2)
Average sample number function	Wald, A. (1947c, p. 25)
Balanced incomplete blocks	Fisher, R.A. and Yates, F. (1938, p. vii)
—— —— ——, partially	Bose, R.C. and Nair, K.R. (1939, title)
Bar chart	Brinton, W.C. (1914, p. 229)
Bayes estimate	Wald, A. (1950, p. 143)
—— factor	Good, I.J. (1958, p. 803)
—— procedure	Wurtele, Z.S. (1949, p. 469)
—— risk	Hodges, J.L. and Lehmann, E.L.

—— solution | Wald, A. (1947a, p. 549)
Bayes's theorem | Lubbock, I.W. and Drinkwater-Bethune, J.E. (c. 1830, p. 48)[2]

Bayesian | Fisher, R.A. (1950, p. 1.2b)[3]
Bell-shaped curve | Galton, F. (1876, p. 14)
Bernoulli trials | Uspensky, J.V. (1937, p. 409)
Best critical region (BCR) | Neyman, J. and Pearson, E.S. (1933a, p. 297)

—— linear unbiased estimate | David, F.N. and Neyman, J. (1938, p. 106)

—— unbiased quadratic estimate | Hsu, P.L. (1938, title)
$\beta_1(= \mu_3^2/\mu_2^3)$, $\beta_2(= \mu_4/\mu_2^2)$ | Pearson, K. (1895, p. 351)
Beta distribution | Pearson, E.S. (1941, p. 152)
 (distribuzione β) | Gini, C. (1911, p. 16)[3]
Biased (errors) | Bowley, A.L. (1897, p. 859)
Bimodal | Williams, S.R. (1903, p. 302)
Binomial distribution | Yule, G.U. (1911, p. 287)[4]
Bioassay | Wood, H.C. (1912, title)
Biometry | Whewell, W. (1831)
Biostatistics | *Webster's Dictionary* (1890)
Bivariate (normal) | Pearson, K. (1920, p. 37)
Bonferroni inequalities | Feller, W. (1950, p. 75)
Bootstrap | Efron, B. (1979, title)
Box plot | Tukey, J.W. (1970, ch. 5)
Branching processes | Kolmogorov, A.N. and Dmitriev, N.A. (1947, title)

Breakdown point | Hampel, F.R. (1971, p. 1894)
Brownian motion | Ramsay, W. (1892)[5]

Canonical correlation | Hotelling, H. (1936, p. 321)
Cauchy distribution | Uspensky, J.V. (1937, p. 275)
 (loi de Cauchy) | Lévy, p. (1925, p. 179)
Censoring | Hald, A. (1949, p. 119)[6]
——, Type I, Type II | Gupta, A.K. (1952, p. 260)
Central limit theorem | Cramér, H. (1937, p. 49)
 (zentraler Grenzwertsatz) | Pólya, G. (1920, title)
Change-over design | Cochran, W.G. et al. (1941, title)
Characteristic function | Kullback, S. (1934, title)
 (fonction caractéristique) | Poincaré, H. (1912, p. 206)[7]
Chi-squared (χ^2) | Pearson, K. (1900, p. 339)
Cluster analysis | Tryon, R.C. (1939, title)
—— sampling | Hansen, M.H. and Hurwitz, W.N. (1943, p. 333)

Coefficient of correlation | Pearson, K. (1896, p. 253)[8]

(1952, p. 399)

—— —— variation | Pearson, K. (1896, p. 253)

Column effect | Wilks, S.S. (1943, p. 187)

Competing risks | Neyman, J. (1950, p. 69)

Complete class (of admissible decision functions) | Wald, A. (1947a, title)

Completeness (of a family of measures) | Lehmann, E.L. and Scheffé, H. (1950, title)

——, bounded | Lehmann, E.L. and Scheffé, H. (1950, p. 306)

Components of variance | Daniels, H.E. (1939, title)

Composite hypothesis | Neyman, J. and Pearson, E.S. (1933a, p. 293)

Computer-intensive | Diaconis, P. and Efron, B. (1983, title)

Confidence coefficient | Neyman, J. (1934, p. 562)

—— interval | Neyman, J. (1934, p. 562)

Confounding | Fisher, R.A. (1926, p. 513)

——, partial | Fisher, R.A. (1935a, p. 131)

Conjugate prior distributions | Raiffa, H. and Schlaifer, R. (1961, p. 47)

Consistency | Fisher, R.A. (1922a, p. 309)

—— (of a test) | Wald, A. and Wolfowitz, J. (1940, p. 153)

Consumer's risk | Dodge, H.F. and Romig, H.G. (1929, p. 614)

Contagious distribution | Neyman, J. (1939, title)

Contingency table | Pearson, K. (1904, p. 474)[9]

Control chart | Shewhart, W. (1931, p. 290)

Convolution | Winter, A. (1934, title)

Correlated | Galton, F. (1875, p. 74)

Correlation | Galton, F. (1888, title)

——, canonical | Hotelling, H. (1936, p. 321)

——, multiple | Pearson, K. (1908, p. 59)

——, partial | Galton, F. (1888, p. 144)

Correlation coefficient | Pearson, K. (1896, p. 253)[10]

—— ——, multiple | Pearson, K. (1914, p. 182)

—— ——, partial | Yule, G.U. (1907, p. 186)

Correlogram | Wold, H. (1938, p. 135)

Covariance | Fisher, R.A. (1930, p. 195)[3]

——, analysis of | Bailey, A.L. (1931, title)

Cox's regression model | Kalbfleisch, J.D. and Prentice, R.L. (1973, title)

Cramér–Rao inequality | Neyman, J. and Scott, E.L. (1948, p. 21)

Critical function | Lehmann, E.L. and Stein, C. (1948,

	p. 512)
Critical region	Neyman, J. and Pearson, E.S. (1933a, p. 289)
—— ——, best	Neyman, J. and Pearson, E.S. (1933a, p. 289)
—— ——, unbiased	Neyman, J. and Pearson, E.S. (1936, p. 8)
Cumulant	Fisher, R.A. and Wishart, J. (1931, p. 198)[5]
Cumulative distribution function (cdf)	Wilks, S.S. (1943, p. 5)
—— sum (control chart)	Page, E.S. (1954, p. 103)
Cyclic (design)	Bose, R.C. and Shimamoto, T. (1952, p. 164)
Decile	Galton, F. (1882, p. 245)
Decision function, statistical	Wald, A. (1945a, title)
—— theory	Ghosh, M.N. (1952, title)
Degrees of freedom	Fisher, R.A. (1922b, p. 88)
—— —— —— of a composite hypothesis	Neyman, J. and Pearson, E.S. (1933a, p. 289)
Deviance	Nelder, J.A. and Wedderburn, R.W.M. (1972, p. 374)
Deviate (normal)	Galton, F. (1907, p. 400)
Discriminant function	Fisher, R.A. (1936b, p. 179)
Dispersion	Galton, F. (1876, p. 13)
——, measures of	Bowley, A.L. (1907, p. 136)
Distribution-free (tolerance limits)	Wilks, S.S. (1943, p. 94)
Distribution function (Verteilungsfunktion)	Doob, J.L. (1935, p. 160)
	Mises, R. v. (1919, p. 67)
——, cumulative (cdf)	Wilks, S.S. (1943, p. 5)
Domain of attraction (domaine d'attraction)	Lévy, P. (1925, p. 252)
Double exponential (Laplace)	Fisher, R.A. (1920, p. 770)
—— sampling	Dodge, H.F. and Romig, H.G. (1929, p. 619)
Dual scaling	Nishisato, S. (1980, title)
Dynamic programming	Bellman, R. (1953, title)
Econometrics	Frish, R. (1933, p. 1)
Efficiency	Fisher, R.A. (1922a, p. 309)
EM algorithm	Dempster, A.P. et al. (1977, title)
Empirical Bayes	Robbins, H. (1956, title)
—— distribution function	Feller, W. (1948, p. 177)
	Kolmogorov, A. (1933, p. 83)

Equivariant	Wijsman, R.A. (1967, p. 391)
Errors of first and second kind	Neyman, J. and Pearson, E.S. (1933a, p. 296)
Estimability	Bose, R.C. (1944, p. 5)
Estimating equation	Yule, G.U. (1902, p. 197)[3]
Estimator	Pitman, E.J.G. (1938, p. 400)
Evolutionary operation	Box, G.E.P. (1957, title)
Exchangeable	Loève, M. (1955, p. 365)[5]
(échangeable)	Fréchet, M. (1939)
Exploratory data analysis	Tukey, J.W. (1970, title)
Exponential (negative exponential)	Pearson, K. (1895, p. 345)
—— (type)	Koopman, B.O. (1936, p. 400)
—— family	Girshick, M.A. and Savage, L.J. (1951, p. 53)
Extreme-value distribution	Lieblein, J. (1953, p. 282)
—— —— theory	Gumbel, E.J. (1954, p. 1)
F	Snedecor, G.W. (1934, p. 15)
Factor analysis	Thurstone, L.L. (1931, p. 406)
Factorial design	Fisher, R.A. (1935a, p. 96)
—— moment	Steffensen, J.F. (1923, title)
Fiducial	Fisher, R.A. (1930a, p. 533)
Fisher-consistent	Rao, C.R. (1957, p. 26)
Fisher information	Fraser, D.A.S. (1968, p. 304)
Fixed effects	Eisenhart, C. (1947, p. 20)
—— model	Scheffé, H. (1956, p. 252)
Fractional replication	Finney, D.J. (1945, title)
Frailty	Vautel, J.W. et al. (1979, title)
Galton–Watson process	Harris, T.E. (1963, p. 2)
Gambler's ruin	Coolidge, J.L. (1909, title)
Game theory	Williams, J.D. (1954, p. vii)[11]
Gamma distribution	Weatherburn, C.E. (1946, p. 149)
Gauss–Markov theorem	Scheffé, H. (1959, p. 14)
Geometric distribution	Feller, W. (1950, p. 174)
Gibbs Sampler	Geman, S. and Geman, D. (1984, p. 722)
Goodness of fit	Pearson, K. (1900, p. 157)
Greco-Latin square	Fisher, R.A. and Yates, F. (1934, p. 493)
Group divisible	Bose, R.C. and Shimamoto, T. (1952, p. 154)
Hat matrix	Hoaglin, D.C. and Welsch, R.E. (1978, title)[12]

Hazard rate — Barlow, R.E. et al. (1963, title)

—— ——, decreasing (DHR) — Barlow, R.E. et al. (1963, p. 378)

—— ——, increasing (IHR) — Barlow, R.E. et al. (1963, p. 377)

Helmert transformation — Lancaster, H.O. (1949, p. 124)

Heteroscedastic — Pearson, K. (1905a, p. 496)[13]

Heteroskedastic — Valavanis, S. (1959, p. 48)[13]

Hierarchical — Fisher, R.A. (1936a, p. 111)

—— Bayes — Good, I.J. (1980, p. 489)

Histogram — Pearson, K. (1895, p. 399)

Homoscedastic — Pearson, K. (1905a, p. 496)[13]

Homoskedastic — Valavanis, S. (1959, p. 50)[13]

Hotelling's T^2 — Simaika, J.B. (1941, p. 70)

Hypothesis

——, alternative — Neyman, J. and Pearson, E.S. (1933a, p. 294)

——, composite, simple — Neyman, J. and Pearson, E.S. (1933a, p. 293)

——, linear — See Linear hypothesis

——, null — Fisher, R.A. (1935a, p. 18)

——, test of (essentially) — Neyman, J. and Pearson, E.S. (1928, p. 175)

Identifiability — Koopmans, T.C. (1949, p. 132)

Importance sampling — Hammersley, J.M. and Handscomb, D.C. (1964, p. 57)

Imputation — Pritzker, L. et al. (1965, p. 442)

Incidence matrix — Connor, W.S., Jr. (1952, p. 60)

Incomplete blocks, balanced — Fisher, R.A. and Yates, F. (1938, p. 10)

—— ——, partially balanced — Bose, R.C. and Nair, K.R. (1939, title)

Index number — Jevons, W.S. (1875, p. 332)

Influence function (curve) — Hampel, F.R. (1974, p. 383)

Information (amount of) — Fisher, R.A. (1925b, p. 709)

—— inequality — Savage, L.J. (1954, p. 238)

—— matrix — Fisher, R.A. (1941b, p. 184)

Interaction — Fisher, R.A. (1926, p. 512)

Interquartile range — Galton, F. (1882, p. 245)

Interval estimation — Mood, A.M. (1950, p. xi)

Invariant — Sylvester, J.J. (1851, p. 396)

Inverse binomial sampling — Tweedie, M.C.K. (1945, p. 453)

—— Gaussian — Tweedie, M.C.K. (1947, p. 47)

J shaped — Yule, G.U. (1911, p. 98)

Jackknife — Miller, R.G. (1964, title)[14]

Kernel estimates	Wegman, E.J. (1972, p. 533)
Kolmogorov–Smirnov statistics	Miller, L.H. (1956, p. 113)
Kriging	Matheron, G. (1963b, p. 1259)
(krigeage)	Matheron, G. (1963a, title)
Kullback–Leibler information	Bahadur, R.R. (1967, p. 13)
Kurtosis	Pearson, K. (1905b, p. 181)
λ	Neyman, J. and Pearson, E.S. (1928, p. 187)
L-estimator	Jaeckel, L.A. (1971, p. 1021)
L-statistic	Boos, D.D. (1979, title)
Lag	Hooker, R.H. (1901, p. 487)
Large deviations	Feller, W. (1950, p. x)
(grosse Abweichungen)	Smirnoff, N. (1933, title)
Latin square	Cayley, A. (1890, title)
(quarré latin)	Euler, L. (1782, p. 90)
Lattice (design)	Yates, F. (1937, p. 85)
Law of iterated logarithm	Hartman, P. and Winter, A. (1941, title)
(Gesetz des iterierten Logarithmus)	Kolmogoroff, A. (1929, title)
Least favorable distribution	Wald, A. (1945a, p. 270)
—— squares	See Method of least squares
—— ——, weighted	Pearson, K. (1920, p. 26)
Level of significance	Fisher, R.A. (1925a, p. 157)
Leverage	Ryan, T.A., Jr. (1978, title)[15]
Likelihood	Fisher, R.A. (1921, p. 24)
——, marginal	Fraser, D.A.S. (1968, p. 188)
——, maximum	Fisher, R.A. (1922a, p. 323)
——, partial	Cox, D.R. (1975, title)
——, profile	Barndorff-Nielsen, O. (1983, p. 35)
——, quasi-	Wedderburn, R.W.M. (1974, title)
Likelihood ratio	Neyman, J. and Pearson, E.S. (1931, p. 480)
—— ——, monotone	Rubin, H. (1951, title)
Linear hypothesis	Kolodziejczyk, S. (1935, p. 161)
—— ——, general	Tang, P.C. (1938, p. 148)
—— ——, generalized	Nelder, J.A. and Wedderburn, R.W.M. (1972, title)
Linear model	Anderson, R.L. and Bancroft, T.A. (1952, p. 169)
—— programming	Dantzig, G.B. (1949, p. 203)
Link function	Nelder, J.A. (1974, p. 327)
Locally best unbiased estimate	Barankin, E.W. (1949, title)

Location — Fisher, R.A. (1922a, p. 310)
—— parameter — Pitman, E.J.G. (1938, title)
lod (log odds) — Barnard, G.A. (1949, p. 116)[3]
Logarithmic series distribution — Kendall, D.G. (1948, title)[16]
Logistic — Verhulst, P.-F. (1845, p. 8)
—— regression — Efron, B. (1975, p. 893)
Logit — Berkson, J. (1944, p. 358)
Log-linear model — Bishop, Y.M.M. and Fienberg, S.E. (1969, p. 119)

Lognormal distribution — Gaddum, J.H. (1945, title)
Loss — Wald, A. (1939, p. 302)
—— function — Hodges, J.L., Jr., and Lehmann, E.L. (1950, p. 182)

M-estimator — Huber, P.J. (1964, p. 74)
Markov chains — Doob, J.L. (1942, title)
  (chaînes de Markoff) — Doeblin, W. (1937, p. 57)
Martingale — Ville, J. (1939, p. 85)
Mathematical statistics — West, C.J. (1918, title)
  (mathematische Statistik) — Knies, C.G.A. (1850, p. 163)[26]
Maximum entropy — Jaynes, E.T. (1957, p. 620)
—— likelihood — Fisher, R.A. (1922a, p. 323)
Mean square (of errors) — Edgeworth, F.Y. (1885, p. 188)
Median — Galton, F. (1882, p. 245)
  (valeur médiane) — Cournot, A.A. (1843, p. 83)[17]
—— absolute deviation — Andrews, D.F. et al. (1972, p. 13)
—— -unbiased — Brown, G.W. (1947, p. 583)
Meta-analysis — Glass, G.V. (1976, title)
Method of least squares — Ivory, J. (1825, title)
  (méthode des moindres quarrés) — Legendre, A.M. (1805, p. viii)
—— maximum likelihood — Fisher, R.A. (1922a, p. 323)
—— moments — Pearson, K. (1902, p. 265)
—— paired comparisons — Thurstone, L.L. (1927, title)
Minimal sufficient statistic — Lehmann, E.L. and Scheffé, H. (1950, p. 307)

Minimax (solution, strategy) — Wald, A. (1947b, p. 282)
Minimum chi-squared — Fisher, R.A. (1928b, p. 251)
—— variance unbiased (essentially) — Aitken, A.C. and Silverstone, H. (1942, p. 188)

Mixed model — Mood, A.M. (1950, p. 348)
Mode — Pearson, K. (1895, p. 345)
Model I, II (in ANOVA) — Eisenhart, C. (1947, pp. 9, 15)
Model, components of variance — Mood, A.M. (1950, p. 342)
——, fixed effects — Scheffé, H. (1956, p. 252)
——, linear — Anderson, R.L. and Bancroft, T.A.

	(1952, p. 169)
——, mixed	Mood, A.M. (1950, p. 348)
——, random effects	Scheffé, H. (1956, p. 252)
Moment	Pearson, K. (1893, p. 615)
——, factorial	Steffensen, J.F. (1923, title)
—— generating function	Craig, C.C. (1936, p. 55)
Monte Carlo method	Metropolis, N. and Ulam, S. (1949, title)[5]
Moving average	King, W.I. (1912, p. 168)[18]
Multidimensional scaling	Torgerson, W.S. (1952, title)
Multinomial distribution	Fisher, R.A. (1925b, p. 719)
Multiple correlation	Pearson, K. and Lee, A. (1897, p. 456)[18]
—— —— coefficient	Pearson, K. (1914, p. 182)
—— comparisons	Duncan, D.B. (1951, p. 178)
Multivariate	Pearson, K. (1920, p. 37)
—— analysis	Bartlett, M.S. (1939, title)
—— normal (essentially)	Wishart, J. (1928, title)
Nearest neighbor	Chandrasekar, S. (1943, p. 2)[5]
Negative binomial distribution	Greenwood, M. and Yule, G.U. (1920, p. 274)
Nested	Ganguli, M. (1941, title)[19]
New better than used (NBU)	Marshall, A.W. and Proschan, F. (1972, p. 396)
Neyman–Pearson lemma	Dantzig, G.B. and Wald, A. (1951, title)
Nonadditivity	Cochran, W.G. (1947, p. 35)
Noncentral	Fisher, R.A. (1928a, p. 670)
Noninformative	Raiffa, H. and Schlaifer, R. (1961, p. xxi)
Nonparametric	Wolfowitz, J. (1942, p. 264)
—— regression	Hanson, D.L. and Pledger, G. (1976, p. 1038 fn)
Nonresponse	Deming, W.E. (1944, p. 360)
Nonsampling error	Hansen, M.H. et al. (1953, p. xv)
Normal (distribution)	Peirce, C.S. (1873, p. 206)[20]
—— equations (Normalgleichungen)	Gauss, C.F. (1822, c. 82)
—— score	Fisher, R.A. and Yates, F. (1938, p. 50)
Nuisance parameter	Hotelling, H. (1940, title)
Null hypothesis	Fisher, R.A. (1935a, p. 18)
Odds ratio	Cox, D.R. (1958, p. 222)[18]
Operating characteristic function	Wald, A. (1945b, p. 162)

Order statistics (essentially)   Wilks, S.S. (1942, p. 403)
—— ——, concomitants of   David, H.A. (1973, title)
—— ——, induced   Bhattacharya, P.K. (1974, title)
Orthogonal array   Bose, R.C. (1950, title)
Orthogonality (in design)   Yates, F. (1933, title)

P value   Deming, W.E. (1943, p. 30)
Paired comparisons, method of   Thurstone, L.L. (1927, title)
Parameter   Kapteyn, J.C. (1903, p. 18)[2]
Pareto distribution   Pigou, A.C. (1920, p. 693)
Partial correlation   Galton, F. (1888, p. 144)
—— —— coefficient   Yule, G.U. (1907, p. 186)
—— regression   Yule, G.U. (1897, p. 833)[21]
Partially balanced incomplete   Bose, R.C. and Nair, K.R. (1939,
    blocks   title)
Path coefficient   Wright, S. (1921, p. 557)
Peakedness   Birnbaum, Z.W. (1948, title)
Pearson type curve (essentially)   Student (1908, p. 4)
Penalized likelihood   Montricher, G.F. de, Tapia, R.A., and
    Thompson, J.R. (1975, p. 1329)
Percentage point (essentially)   Fisher, R.A. (1925a, p. 198)
Percentile   Galton, F. (1885a, p. 276)
Periodogram   Schuster, A. (1898, p. 24)
Permutation test   Box, G.E.P. and Andersen, S.L.
    (1955, p. 3)
Petersburg paradox   Keynes, J.M. (1921, p. 316)[5]
Pie chart   Haskell, A.C. (1922, p. xiv)
Pivotal quantity   Fisher, R.A. (1941a, p. 147)
Point estimation   Wilks, S.S. (1943, p. 122)
Poisson distribution (essentially)   Soper, H.E. (1914, title)
Poisson process   Feller, W. (1949, p. 405)
Posterior probability   Wrinch, D. and Jeffreys, H. (1921,
    p. 387)[22]
Power (of a test)   Neyman, J. and Pearson, E.S.
    (1933b, p. 498)
—— function   Neyman, J. and Pearson, E.S. (1936,
    p. 20)
Principal components   Hotelling, H. (1933, title)
Prior probability   Wrinch, D. and Jeffreys, H. (1921,
    p. 381)[22]
Probability density   Jeffreys, H. (1939, p. 24)
    (Wahrscheinlichkeitsdichte)   Markoff, A.A. (1912, p. 155)
Probability density function   Wilks, S.S. (1943, p. 8)
—— function   Aitken, A.C. (1939, p. 16)
—— generating function   Seal, H.L. (1949, p. 67)[23]

—— mass function   Parzen, E. (1960, p. 167)
—— paper   Hazen, A. (1914, p. 1549)
Probit   Bliss, C.I. (1934, title)
—— analysis   Finney, D.J. (1944, title)
Producer's risk   Dodge, H.F. and Romig, H.G. (1941,
                            p. 7)
Product-limit estimate   Kaplan, E.L. and Meier, P. (1958,
                            p. 457)
Projection pursuit   Friedman, J.H. and Tukey, J.W.
                            (1974, title)

Q–Q plot   Wilk, M.B. and Gnanadesikan, R.
                            (1968, p. 1)
Quantile   Kendall, M.G. (1940, title)
—— function   Parzen, E. (1979, p. 105)
Quartile   McAlister, D. (1879, p. 374)
—— lower   McAlister, D. (1879, p. 374)
—— upper   Galton, F. (1882, p. 245)

ρ Edgeworth, F.Y. (1892, p. 190)
Random effects   Eisenhart, C. (1947, p. 20)
—— model   Scheffé, H. (1956, p. 252)
—— number   Pearson, K. (1927, p. iii)
—— sampling   Pearson, K. (1900, title)
—— variable   Winter, A. (1934, p. 660)
  (variabile casuale)   Cantelli, F.P. (1916, p. 192)[17]
—— walk   Pearson, K. (1905c, title)
Randomization   Fisher, R.A. (1926, p. 510)
—— tests   Box, G.E.P. and Andersen, S.L.
                            (1955, p. 3)
—— theory   Pitman, E.J.G. (1937, p. 322)
Randomized blocks   Fisher, R.A. (1926, p. 509)
—— response   Warner, S.L. (1965, title)
Range   Lloyd, H. (1848, p. 182)
——, quasi   Mosteller, F. (1946, p. 391)
Ranked-set sampling   Halls, L.K. and Dell, T.R. (1966,
                            title)
Rankit   Ipsen, J. and Jerne, N.K. (1944,
                            p. 349)
Rao–Blackwellization   Berkson, J. (1955, p. 142)
Ratio estimate   Deming, W.E. (1950, p. xii)
Receiver operating characteristic   Pollack, I. and Decker, L.R. (1958,
  (ROC)   title)
Recovery of interblock information   Yates, F. (1939, title)
Regression   Galton, F. (1885b, p. 507)

——, partial   Yule, G.U. (1897, p. 833)
Resampling   Efron, B. (1979, p. 1)
Resistance   Andrews, D.F. (1974, p. 523)
Resolvable design   Bose, R.C. (1942, p. 105)
Response surface   Box, G.E.P. and Wilson, K.B. (1951,
                                         p. 2)
Ridge regression   Hoerl, A.E. and Kennard, R.W.
                                         (1970, title)
Risk, risk function   Wald, A. (1939, p. 304)
Robustness   Box, G.E.P. (1953, p. 318)
Rotatable design   Box, G.E.P. and Hunter, J.S. (1957,
                                         p. 195)
Roughness penalty   Good, I.J. (1971, title)
Row effect   Wilks, S.S. (1943, p. 187)

σ Pearson, K. (1894, p. 80)
Sampling distribution   Fisher, R.A. (1928, title)
Scale   Fisher, R.A. (1922a, p. 337)
—— parameter   Pitman, E.J.G. (1938, p. 391)
Scaling   Fisher, R.A. (1922a, p. 310)
Scatterplot   Kurtz, A.K. and Edgerton, H.A.
                                         (1939, p. 151)
Score (ideal)   Fisher, R.A. (1935b, p. 193)
——, efficient   Rao, C.R. (1948, p. 51)
——, normal   Fisher, R.A. and Yates, F. (1938,
                                         p. 50)
Semiparametric   Kalbfleisch, J.D. (1978,
                                         p. 214)
Sequential analysis   Wald, A. (1945b, p. 120)
—— probability ratio test   Wald, A. (1945b, p. 125)
—— test   Wald, A. (1945b, title)
Serial correlation   Yule, G.U. (1926, p. 14)
Sheppard's corrections   Pearson, K. (1901, p. 451)
Shrinkage   Thompson, J.R. (1968)
Sign test   Stewart, W.M. (1941, title)
  (Vorzeichenprüfung)   Helmert, F.R. (1905, p. 594)
Signed rank test   See Section 2
Significance   Edgeworth, F.Y. (1885, p. 182)
——, level of   Fisher, R.A. (1925a, p. 43)
——, test of   Fisher, R.A. (1925a, p. 157)
Similar region   Neyman, J. and Pearson, E.S.
                                         (1933a, p. 312)
Simple hypothesis   Neyman, J. and Pearson, E.S.
                                         (1933a, p. 312)
—— random sampling   Cochran, W.G. (1953, p. 11)

Size (of critical region)	Neyman, J. and Pearson, E.S. (1933a, p. 313)
Skewness	Pearson, K. (1895, p. 357)
Slippage test	Mosteller, F. (1948, title)
Split plot	Yates, F. (1935, p. 197)
Stable law (loi stable)	Lévy, P. (1923, title)
Standard deviation (σ)	Pearson, K. (1894, p. 80)
—— error	Yule, G.U. (1897, p. 821)
Stationary processes (stationäre stochastische Prozesse)	Cramér, H. (1947, p. 188) Khintchine, A. (1934, title)
Statistic	Fisher, R.A. (1921, p. 5)[2]
Statistical decision function	Wald, A. (1945a, title)
—— science	Anonymous (1838, p. 1)[3]
Statistics	See Section 3
——, mathematical	See Mathematical statistics
Stem-and-leaf display	Tukey, J.W. (1972, p. 295)
Stochastic processes (stochastische Prozesse)	Doob, J.L. (1934, title) Khintchine, A. (1934, title)
Stochastically larger	Mann, H.B. and Whitney, D.R. (1947, title)
Student's t (essentially)	Fisher, R.A. (1924, p. 809)
Studentization	Fisher, R.A. (1934b, p. 619)[25]
——, internal, external	David, H.A. (1970, p. 70)
Studentized range	Pearson, E.S. and Hartley, H.O. (1943, title)
Subjective probability	See Daston, L. (1994)[26]
Sufficiency	Fisher, R.A. (1922a, p. 310)
Sufficient statistic	Fisher, R.A. (1925b, p. 712)
Superefficiency	LeCam, L. (1953, p. 281)
Survival function	Kaplan, E.L. and Meier, P. (1958, p. 461)
t	Fisher, R.A. (1924, p. 809)
Test criterion	Neyman, J. and Pearson, E.S. (1928, title)
—— of hypothesis	Neyman, J. and Pearson, E.S. (1928, p. 175)
—— of significance	Fisher, R.A. (1925a, p. 43)
Time series	Persons, W.M. (1919, p. 123)
Tolerance limits (statistical)	Wilks, S.S. (1941, title)
Treatment effect	Wilks, S.S. (1943, p. 187)
Trend	Hooker, R.H. (1901, p. 486)
Trimmed mean	Tukey, J.W. (1962, p. 18)
Trimming	Tukey, J.W. (1962, p. 1)

Truncation	Pearson, K. and Lee, A. (1908, p. 63)
2 × 2 table	Barnard, G.A. (1945, title)
Type I and Type II errors	Neyman, J. and Pearson, E.S. (1933b, p. 496)
U-shaped	Yule, G.U. (1911, p. 102)
U-statistic	Hoeffding, W. (1948, p. 293)
Unbiased (errors)	Bowley, A.L. (1897, p. 859)
—— critical region	Neyman, J. and Pearson, E.S. (1936, p. 8)
Uniform distribution	Uspensky, J.V. (1937, p. 239)
Uniformly minimum variance unbiased	Lehmann, E.L. and Stein, C. (1950, p. 377)
—— most powerful test	Neyman, J. and Pearson, E.S. (1936, p. 1)
Unimodal	Helguero, F. de (1904, p. 84)
Variance	Fisher, R.A. (1918a, p. 399)
——, analysis of	Fisher, R.A. (1918b, p. 219)
——, components of	Daniels, H.E. (1939, title)
—— function	Finney, D.J. and Phillips, P. (1977, title)
Variate	Pearson, K. (1909, p. 97)
—— difference method	Cave, B.M. and Pearson, K. (1914, title)
Variogram	Matheron, G. (1963b, p. 1250)
Weibull distribution	Lieblein, J. (1955, title)
Weighted least squares	Pearson, K. (1920, p. 26)[5]
Winsorized	Dixon, W.J. (1960, p. 385)
Yates's correction for continuity	Fisher, R.A. (1936a, p. 97)
Youden square	Fisher, R.A. (1938, p. 422)
z-distribution	Fisher, R.A. (1924, p. 496)
Zero-sum game	Neumann, J. von and Morgenstern, O. (1944, p. ii)

Notes

[1]Term (and concept) introduced by E.J.G. Pitman in 1948 in lectures at Columbia University; also in Pitman's mimeographed lecture notes on Non-Parametric Statistical Inference at the University of North Carolina.

[2]Reference supplied by A. Hald.

[3]Reference supplied by A.W.F. Edwards.

[4]Even in a probabilistic context the use of "binomial," alone or in other combinations, such as "binomial law," goes back much further.

[5]See also Section 3.

[6]Hald states that the term was suggested to him by J.E. Kerrich.

[7]Poincaré's term actually means today's "moment generating function."

[8]Edgeworth (1892, p. 191) uses the term for a different estimate of ρ.

[9]Page number in K. Pearson (1956).

[10]See Stigler (1986) for an excellent discussion of the history of correlation.

[11]See also David (1998, p. 36).

[12]The authors attribute the term to J.W. Tukey.

[13]See also Section 4.

[14]Term due to Tukey (unpublished).

[15]Author reports that the word "was floating around at the time."

[16]Williams (1944) uses just "logarithmic series."

[17]Given in Sheynin (1997).

[18]Reference supplied by A.M. Hughes.

[19]Author attributes term to P.C. Mahalanobis.

[20]See also Section 2.

[21]Here "partial regression" is a second choice to "net regression," but in Yule (1907) only "partial" survives.

[22]A. Hald informs me that "probability a posteriori" and "p. a priori" occur in Lubbock and Drinkwater-Bethune (c. 1830, p. 10 and p. 25).

[23]Uspensky (1937) treats "generating functions of probability." The concept is, of course, very much older and goes back at least to de Moivre.

[24]Term first given in Wald (1943).

[25]Term due to Gosset (1932).

[26]Reference supplied by O. Sheynin.

5 Acknowledgments

The list of first(?) occurrences in this appendix represents a merging of two earlier lists in David (1995, 1998), together with additional terms and improvements in dates of first occurrence. Acknowledgments in the two previous papers make clear the debt I owe many correspondents, especially Anthony Edwards, Jack Good, Oliver Lancaster, and Erich Lehmann.

Anthony Edwards has continued to provide valuable help for this appendix. New multiple contributors are Anders Hald and Alan Hughes. All three are repeatedly acknowledged on the list itself for the references they supplied. I am also grateful to David Finney, Friedrich Pukelsheim, Hans Riedwyl, Shayle Searle, Oscar Sheynin, Stephen Stigler, and John Stufken for suggestions and other help specifically improving this appendix.

References

For this appendix only, the following journal abbreviations are used.

AE	*Annals of Eugenics*
AHES	*Archive for the History of the Exact Sciences*
AMS	*Annals of Mathematical Statistics*
AS	*Annals of Statistics*
ApS	*Applied Statistics*
CRAS	*Comptes rendus de l'académie des sciences de Paris*
JASA	*Journal of the American Statistical Association*
JRSS A,B	*Journal of the Royal Statistical Society*, Series A, B
PBS2	*Proceedings of the Second Berkeley Symposium on Mathematical Statistics and Probability*
PCPS	*Proceedings of the Cambridge Philosophical Society*
PLMS	*Proceedings of the London Mathematical Society*
PRS A	*Proceedings of the Royal Society*, Series A.
PRSE A	*Proceedings of the Royal Society of Edinburgh*, Series A
PRSL A	*Proceedings of the Royal Society of London*, Series A
PTRSL A	*Philosophical Transactions of the Royal Society of London*, Series A
SA	*Skandinavisk Aktuarietidskrift*
SRM	*Statistical Research Memoirs*
TAS	*The American Statistician*
TRSE	*Transactions of the Royal Society of Edinburgh*

Aitken, A.C. (1939). *Statistical Mathematics*. Oliver and Boyd, Edinburgh.

Aitken, A.C. and Silverstone, H. (1942). On the estimation of statistical parameters. *PRSE* A, **61**, 186–194.

Anderson, R.L. and Bancroft, T.A. (1952). *Statistical Theory in Research*. McGraw-Hill, New York.

Andrews, D.F. (1974). A robust method for multiple linear regression. *Technometrics*, **16**, 523–531.

Andrews, D.F., Bickel, P.J., Hampel, F.R., Huber, P.J., Rogers, W.H., and Tukey, J.W. (1972). *Robust Estimates of Location*. Princeton University Press.

Anonymous (1838). Introduction. *J. Statist. Soc. London*, **1**, 1–5.

Bahadur, R.R. (1967). An optimal property of the likelihood ratio statistic. *PBS5*, Vol. 1, 13–26.

Bailey, A.L. (1931). The analysis of covariance. *JASA*, **26**, 424–435.

Barankin, E.W. (1949). Locally best unbiased estimates. *AMS*, **20**, 477–501.

Barlow, R.E. (1991). Introduction to de Finetti (1937) Foresight: Its logical laws, its subjective sources. In S. Kotz and N.L. Johnson (eds.), *Breakthroughs in Statistics*, Vol. I. Springer, New York, pp. 127–133.

Barlow, R.E., Marshall, A.W., and Proschan, F. (1963). Properties of probability distributions with monotone hazard rate. *AMS*, **34**, 375–389.

Barnard, G.A. (1945). A new test for 2 × 2 tables. *Nature*, **156**, 177.

Barnard, G.A. (1949). Statistical inference. *JRSS* B, **11**, 115–139.

Barndorff-Nielsen, O. (1983). On a formula for the distribution of the maximum likelihood estimator. *Biometrika*, **70**, 343–365.

Bartlett, M.S. (1939). A note on tests of significance in multivariate analysis. *PCPS*, **35**, 180–185.

Bellman, R. (1953). Dynamic programming and a new formalism in the calculus of variations. *Proc. Nat. Acad. Sci.*, **39**, 1077–1082.

Berkson, J. (1942). Tests of significance considered as evidence. *JASA*, **37**, 325–335.

Berkson, J. (1944). Application of the logistic function to bio-assay. *JASA*, **39**, 357–365.

Berkson, J. (1955). Maximum likelihood and minimum χ^2 estimates of the logistic function. *JASA*, **50**, 130–162.

Bhattacharya, P.K. (1974). Convergence of sample paths of normalized sums of induced order statistics. *AS*, **2**, 1034–1039.

Birnbaum, Z.W. (1948). On random variables with comparable peakedness. *AMS*, **19**, 76–81.

Bishop, Y.M.M. and Fienberg, S.E. (1969). Incomplete two-dimensional contingency tables. *Biometrics*, **25**, 119–128.

Bliss, C.I. (1934). The method of probits. *Science*, **79**, 38–39.

Boos, D.D. (1979). A differential for L-statistics. *AS*, **7**, 955–959.

Bose, R.C. (1942). A note on the resolvability of balanced incomplete block designs. *Sankhyā*, **6**, 105–110.

Bose, R.C. (1944). The fundamental theorem of linear estimation. In *Proc. 31st Indian Science Congress, Part III*, pp. 4–5.

Bose, R.C. (1950). A note on orthogonal arrays. *AMS*, **21**, 304–305.

Bose, R.C. and Nair, K.R. (1939). Partially balanced incomplete block designs. *Sankhyā*, **4**, 337–372.

Bose, R.C. and Shimamoto, T. (1952). Classification and analysis of partially balanced incomplete block designs with two associate classes. *JASA*, **47**, 151–184.

Bowley, A.L. (1897). Relations between the accuracy of an average and that of its constituent parts. *JRSS*, **60**, 855–866.

Bowley, A.L. (1907). *Elements of Statistics*. King, London.

Box, G.E.P. (1953). Non-normality and tests on variances. *Biometrika*, **40**, 318–335.

Box, G.E.P. (1957). Evolutionary operation: A method for increasing industrial productivity. *ApS*, **6**, 81–101.

Box, G.E.P. and Andersen, S.L. (1955). Permutation theory in the derivation of robust criteria and the study of departures from assumption. *JRSS* B, **17**, 1–34.

Box, G.E.P. and Hunter, J.S. (1957). Multi-factor experimental designs for exploring response surfaces. *AMS*, **28**, 195–241.

Box, G.E.P. and Wilson, K.B. (1951). On the experimental attainment of optimum conditions. *JRSS* B, **13**, 1–45.

Brinton, W.C. (1914). Graphic methods for presenting data. IV. Time charts. *Engineering Mag.*, **48**, 229–241.

Brown, G.W. (1947). On small-sample estimation. *AMS*, **18**, 582–585.

Brush, S.G. (1968). A history of random processes. I. Brownian movement from Brown to Perrin. *AHES*, **5**, 1–36. [Reproduced in Kendall and Plackett (1977).]

Cantelli, F.P. (1916). La tendenza ad un limite nel senso del calcolo delle probabilità. *Rendiconti del circolo matematico di Palermo*, **41**, 191–201.

Cave, B.M. and Pearson, K. (1914). Numerical illustrations of the variate difference correlation method. *Biometrika*, **10**, 340–355.

Cayley, A. (1890). On Latin squares. *Messenger Math.*, **19**, 135–137.

Chandrasekar, S. (1943). Stochastic problems in physics and astronomy. *Revs. Mod. Physics*, **15**, 1–89.

Cochran, W.G. (1947). Some consequences when the assumptions for the analysis of variance are not satisfied. *Biometrics*, **3**, 22–38.

Cochran, W.G. (1953). *Sampling Techniques*. Wiley, New York.

Cochran, W.G., Autrey, K.M., and Cannon, C.Y. (1941). A double change-over design for dairy cattle feeding experiments. *J. Dairy Sci.*, **24**, 937–951.

Connor, W.S., Jr. (1952). On the structure of balanced incomplete block designs. *AMS*, **23**, 57–71.

Coolidge, J.L. (1909). The gambler's ruin. *Ann. Math.* Ser. 2, **10**, 181–192.

Cournot, A.A. (1843). *Exposition de la théorie des chances et des probabilités.* Hachette, Paris. [Reprinted 1984, J. Vrin, Paris.]

Cox, D.R. (1958). The regression analysis of binary sequences. *JRSS* B, **20**, 215–242.

Cox, D.R. (1975). Partial likelihood. *Biometrika*, **62**, 269–276.

Craig, C.C. (1936). Sheppard's corrections for a discrete variable. *AMS*, **7**, 55–61.

Cramér, H. (1937). *Random Variables and Probability Distributions, Cambridge Tracts in Mathematics*, **36**.

Cramér, H. (1946). *Mathematical Methods of Statistics*. Princeton University Press.

Cramér, H. (1947). Problems in probability theory. *AMS*, **18**, 165–193.

Czuber, E. (1914). *Wahrscheinlichkeitsrechnung, Vol. 1*. Teubner, Leipzig.

Daniels, H.E. (1939). The estimation of components of variance. *Suppl. JRSS*, **6**, 186–197.

Dantzig, G.B. (1949). Programming of independent activities. II. Mathematical model. *Econometrica*, **17**, 200–211.

Dantzig, G.B. and Wald, A. (1951). On the fundamental lemma of Neyman and Pearson. *AMS*, **22**, 87–93.

Daston, L. (1994). How probabilities came to be objective and subjective. *Historia Mathematica*, **21**, 330–344.

David, F.N. and Neyman, J. (1938). Extension of the Markoff theorem on least squares. *SRM*, **2**, 105–116.

David, H.A. (1970). *Order Statistics*. Wiley, New York.

David, H.A. (1973). Concomitants of order statistics. *Bull. Inst. Intern. Statist.*, **45**(1), 295–300.

David, H.A. (1995). First(?) occurrence of common terms in mathematical statistics. *TAS*, **49**, 121–133.

David, H.A. (1998). First(?) occurrence of common terms in probability and statistics — a second list, with corrections. *TAS*, **52**, 36–40.

Deming, W.E. (1943). *Statistical Adjustment of Data*. Wiley, New York.

Deming, W.E. (1944). On errors in surveys. *Amer. Sociol. Rev.*, **9**, 359–369.

Deming, W.E. (1950). *Some Theory of Sampling*. Wiley, New York.

Dempster, A.P., Laird, N.M., and Rubin, D.B. (1977). Maximum likelihood from incomplete data via the EM algorithm. *JRSS* B, **39**, 1–38.

Diaconis, P. and Efron, B. (1983). Computer-intensive methods in statistics. *Scientific American*, **248**, 5, 116–128.

Dixon, W.J. (1960). Simplified estimation from censored normal samples. *AMS*, **31**, 385–391.

Dodge, H.F. and Romig, H.G. (1929). A method of sampling inspection. *Bell System Tech. J.*, **8**, 613–631.

Dodge, H.F. and Romig, H.G. (1941). Single sampling and double sampling inspection tables. *Bell System Tech. J.*, **20**, 1–61.

Doeblin, W. (1937). Sur les propriétés asymptotiques de mouvements régis par certains types de chaînes simples. *Bull. math. soc. roumaine des sciences*, **39** (1), 57–115.

Doob, J.L. (1934). Stochastic processes and statistics. *Proc. Nat. Acad. Sci.*, **20**, 376–379.

Doob, J.L. (1935). The limiting distributions of certain statistics. *AMS*, **6**, 160–169.

Doob, J.L. (1942). Topics in the theory of Markoff chains. *Trans. Amer. Math. Soc.*, **52**, 37–64.

Duncan, D.B. (1951). A significance test for differences between ranked treatments in an analysis of variance. *Virginia J. Sci.*, **2**, 171–189.

Edgeworth, F.Y. (1885). Methods of Statistics. In *Jubilee Volume, RSS*, pp. 181–217.

Edgeworth, F.Y. (1892). Correlated averages. *Phil. Mag.*, 5th Series, **34**, 190–204.

Efron, B. (1975). The efficiency of logistic regression compared to normal discriminant analysis. *JASA*, **70**, 892–898.

Efron, B. (1979). Bootstrap methods: Another look at the jackknife. *AS*, **7**, 1–26.

Efron, B. and Tibshirani, R.J. (1993). *An Introduction to the Bootstrap*. Chapman and Hall, New York.

Eisenhart, C. (1947). The assumptions underlying the analysis of variance. *Biometrics*, **3**, 1–21.

Euler, L. (1782). Recherches sur une nouvelle espèce de quarrés magiques. *Verhandelingen uitgegeven door het zeeuwsch Genootschap der Wetenschappen te Vlissingen*, **9**, 85–239. [Reproduced in *Leonhardi Euleri opera omnia*. Sub auspiciis societatis scientiarium naturalium helveticae, 1st series, vol. 7, pp. 291–392.]

Feller, W. (1948). On the Kolmogorov–Smirnov limit theorems for empirical distributions. *AMS*, **19**, 177–189.

Feller, W. (1949). On the theory of stochastic processes, with particular reference to applications. *PBS*, pp. 403–432.

Feller, W. (1950). *An Introduction to Probability Theory and Its Applications*. Wiley, New York.

Finetti, B. de (1930). Funzione caratteristica di un fenomeno aleatorio. *Memorie della reale accademia dei Lincei, classe di scienze, fisiche, matematiche e naturali*, **4**, 86–133.

Finney, D.J. (1944). The application of probit analysis to the results of mental tests. *Psychometrika*, **9**, 31–39.

Finney, D.J. (1945). The fractional replication of factorial arrangements. *Ann. Eugen.*, **12**, 291–301.

Finney, D.J. and Phillips, P. (1977). The form and estimation of a variance function, with particular reference to radioimmunoassay. *ApS*, **26**, 312–320.

Fisher, R.A. (1918a). The correlation between relatives on the supposition of Mendelian inheritance. *TRSE*, **52**, 399–433.

Fisher, R.A. (1918b). The causes of human variability. *Eugen. Rev.*, **10**, 213–220.

Fisher, R.A. (1920). A mathematical examination of determining the accuracy of an observation by the mean error, and by the mean square error. *Monthly Notices Roy. Astron. Soc.*, **80**, 758–770.

Fisher, R.A. (1921). On the "probable error" of a coefficient of correlation deduced from a small sample. *Metron*, **1**, 4, 3–32.

Fisher, R.A. (1922a). On the mathematical foundations of theoretical statistics. *PTRSL* A, **222**, 309–368.

Fisher, R.A. (1922b). On the interpretation of χ^2 from contingency tables, and the calculation of *P*. *JRSS*, **85**, 87–94.

Fisher, R.A. (1924). On a distribution yielding the error functions of several well known statistics. In *Proc. Intern. Congress Math.*, Toronto, **2**, 805–813.

Fisher, R.A. (1925a). *Statistical Methods for Research Workers*. Oliver and Boyd, Edinburgh.

Fisher, R.A. (1925b). Theory of statistical estimation. *PCPS*, **22**, 700–725.

Fisher, R.A. (1926). The arrangement of field experiments. *J. Ministry Agric., Great Britain*, **33**, 503–513.

Fisher, R.A. (1928a). The general sampling distribution of the multiple correlation coefficient. *PRSL* A, **121**, 654–673.

Fisher, R.A. (1928b). *Statistical Methods for Research Workers*, 2nd edn. Oliver and Boyd, Edinburgh.

Fisher, R.A. (1929). Moments and product moments of sampling distributions. *PLMS*, Ser. 2, **30**, 3, 199–238.

Fisher, R.A. (1930a). Inverse probability. *PCPS*, **26**, 528–535.

Fisher, R.A. (1930b). *The Genetical Theory of Natural Selection*. Oxford University Press.

Fisher, R.A. (1934a). Two new properties of mathematical likelihood. *PRSL* A, **144**, 285–307.

Fisher, R.A. (1934b). Discussion of Neyman (1934).

Fisher, R.A. (1935a). *The Design of Experiments*. Oliver and Boyd, Edinburgh.

Fisher, R.A. (1935b). The detection of linkage with 'dominant' abnormalities. *AE*, **6**, 187–201.

Fisher, R.A. (1936a). *Statistical Methods for Research Workers*, 6th edn. Oliver and Boyd, Edinburgh.

Fisher, R.A. (1936b). The use of multiple measurements in taxonomic problems. *AE*, **7**, 179–188.

Fisher, R.A. (1938). The mathematics of experimentation. *Nature*, **142**, 442–443.

Fisher, R.A. (1941a). The asymptotic approach to Behrens's integral, with further tables for the d test of significance. *AE*, **11**, 141–172.

Fisher, R.A. (1941b). The negative binomial distribution. *AE*, **11** 182–187.

Fisher, R.A. (1950). *Contributions to Mathematical Statistics*. Wiley, New York.

Fisher, R.A. (1971). *Collected Papers of R.A. Fisher*. J.H. Bennett (ed.), University of Adelaide, Australia.

Fisher, R.A. and Wishart, J. (1931). The derivation of the pattern formulae of two-way partitions from those of simpler patterns. *PLMS*, Ser. 2, **33**, 195–208.

Fisher, R.A. and Yates, F. (1934). The 6×6 Latin squares. *PCPS*, **30**, 492–507.

Fisher, R.A. and Yates, F. (1938). *Statistical Tables for Biological, Agricultural and Medical Research*. Oliver and Boyd, London.

Fisher, R.A., Corbet, A.S., and Williams, C.B. (1943). The relation between the number of species and the number of individuals in a random sample of an animal population. *J. Animal Ecology*, **12**, 42–58.

Fraser, D.A.S. (1968). *The Structure of Inference*. Wiley, New York.

Friedman, J.H. and Tukey, J.W. (1974). A projection pursuit algorithm for exploratory data analysis. *IEEE Trans. on Computers*, C-**23**, 881–890.

Frisch, R. (1933). Editorial. *Econometrica*, **1**, 1–2.

Fry, T.C. (1928). *Probability and Its Engineering Uses*. Van Nostrand, New York.

Gaddum, J.H. (1945). Lognormal distributions. *Nature*, **156**, 463–466.

Galton, F. (1875). *English Men of Science*. Appleton, New York.

Galton, F. (1876). *Catalogue of the Special Loan Collection of Scientific Apparatus at the South Kensington Museum*. Her Majesty's Stationery Office, London.

Galton, F. (1882). Report of the Anthropometric Committee. In *Rep. 51st Meeting British Association for the Advancement of Science*, 1881, pp. 245–260.

Galton, F. (1885a). Some results of the Anthropometric Laboratory. *J. Anthrop. Inst.*, **14**, 275–287.

Galton, F. (1885b). Section H. Anthropology. Opening address by Francis Galton. *Nature*, **32**, 507–510.

Galton, F. (1888). Co-relations and their measurement. *PRSL*, **45**, 135–145.

Galton, F. (1889). *Natural Inheritance*. Macmillan, London.

Galton, F. (1907). Grades and deviates. *Biometrika*, **5**, 400–406.

Ganguli, M. (1941). A note on nested sampling. *Sankhyā*, **5**, 449–452.

Gauss, C.F. (1822). Anwendung der Wahrscheinlichkeitsrechnung auf eine Aufgabe der practischen Geometrie. *Astron. Nachr.*, **6**, 81–86.

Geman, S. and Geman, D. (1984). Stochastic relaxation, Gibbs distributions, and the Bayesian restoration of images. *IEEE Trans. on Pattern Analysis and Machine Intelligence*, **6**, 721–741.

Ghosh, M.N. (1952). An extension of Wald's decision theory to unbounded weight functions. *Sankhyā*, **12**, 8–26.

Gibbons, J.D. and Pratt, J.W. (1975). *P*-values: interpretation and methodology. *TAS*, **29**, 20–25.

Gini, C. (1911). Considerazioni sulle probabilità posteriori e applicazioni al rapporto dei sessi nelle nascite umane. Studi economico-giuridici della Università dc Cagliari, Anno III, 5–41 [Reproduced in *Metron* **15**, 133–171 (1949).]

Girshick, M.A. and Savage, L.J. (1951). Bayes and minimax estimates for quadratic loss functions. *PBS2*, 53–73.

Glass, G.V. (1976). Primary, secondary and meta-analysis of research. *Educational Researcher*, 5, **10**, 3–8.

Good, I.J. (1958). Significance tests in parallel and in series. *JASA*, **53**, 799–813.

Good, I.J. (1971). Nonparametric roughness penalty for probability densities. *Nature, Phys. Sci.*, **229**, 29–30.

Good, I.J. (1980). Some history of the hierarchical Bayesian methodology. *Trabajos de estadistica*, **31**, 489–504.

Gosset, W.S. (1932). Letter to E.S. Pearson. In E.S. Pearson (1939, p. 246).

Greenwood, M. and Yule, G.U. (1920). An inquiry into the nature of frequency distributions. *JRSS*, **83**, 255–279.

Gumbel, E.J. (1954). *Statistical Theory of Extreme Values and Some Prac-*

tical Applications. National Bureau of Standards, Applied Mathematics Series, 33.

Gupta, A.K. (1952). Estimation of the mean and standard deviation of a normal population from a censored sample. *Biometrika,* **39,** 260–273.

Hald, A. (1949). Maximum likelihood estimation of the parameters of a normal distribution which is truncated at a known point. *SA,* **1949,** 119–134.

Halls, L.K. and Dell, T.R. (1966). Trial of ranked-set sampling for forage yields. *Forest Sci.,* **12,** 22–26.

Halperin, M. (1960). Some asymptotic results for a coverage problem. *AMS,* **31,** 1063–1076.

Hammersley, J.M. and Handscomb, D.C. (1964). *Monte Carlo Methods.* Methuen, London; Wiley, New York.

Hampel, F.R. (1971). A general qualitative definition of robustness. *AMS,* **42,** 1887–1896.

Hampel, F.R. (1974). The influence curve and its role in robust estimation. *JASA,* **69,** 383–393.

Hansen, M.H. and Hurwitz, W.N. (1943). On the theory of sampling from finite populations. *AMS,* **14,** 333–362.

Hansen, M.H., Hurwitz, W.N., and Madow, W.G. (1953). *Sample Survey Methods and Theory,* Vol. 1. Wiley, New York.

Hanson, D.L. and Pledger, G. (1976). Consistency in concave regression. *AS,* **4,** 1038–1050.

Harris, T.E. (1963). *The Theory of Branching Processes.* Springer, New York.

Harter, H.L. (1978). *A Chronological Annotated Bibliography on Order Statistics, Vol. 1: Pre-1950.* U.S. Government Printing Office, Washington, DC.

Hartman, P. and Winter, A. (1941). On the law of the iterated logarithm. *Amer. J. Math.,* **63,** 169–176.

Haskell, A.C. (1922). *Graphic Charts in Business.* Codex, New York.

Hazen, A. (1914). Storage to be provided in impounding reservoirs for municipal water supply. *Trans. Amer. Soc. Civil Engineers,* **77,** 1539–1669.

Helguero, F. de (1904). Sui massimi delle curve dimorfiche. *Biometrika,* **3,** 84–98.

Helmert, F.R. (1905). Über die Genauigkeit der Kriterien des Zufalls bei Beobachtungsreihen. *Sitz.-Ber. Kgl. Preuss. Akad. Wiss.,* Hlbbd 1, 594–612.

Hoaglin, D.C. and Welsch, R.E. (1978). The hat matrix in regression and ANOVA. *TAS,* **32,** 17–22.

Hodges, J.L., Jr. and Lehmann, E.L. (1950). Some problems in minimax point estimation. *AMS,* **21,** 182–192.

Hodges, J.L. and Lehmann, E.L. (1952). The use of previous experience in reaching statistical decisions. *AMS,* **23,** 396–407.

Hoeffding, W. (1948). A class of statistics with asymptotically normal distribution. *AMS*, **19**, 293–325.

Hoerl, A.E. and Kennard, R.W. (1970). Ridge regression: Applications to nonorthogonal problems. *Technometrics*, **12**, 69–82.

Hogg, R.V. and Craig, A.T. (1959). *Introduction to Mathematical Statistics*. Macmillan, New York.

Hooker, R.H. (1901). Correlations of the marriage rate with trade. *JRSS*, **64**, 485–492.

Hotelling, H. (1933). Analysis of a complex of statistical variables into principal components. *J. Educ. Psychol.*, **24**, 417–441.

Hotelling, H. (1936). Relations between two sets of variates. *Biometrika*, **28**, 321–377.

Hotelling, H. (1940). The selection of variates for use in prediction with some comments on the general problem of nuisance parameters. *AMS*, **11**, 271–283.

Hsu, P.L. (1938). On the best unbiased quadratic estimate of the variance. *SRM*, **2**, 91–104.

Huang, J.S. (1972). A note on Robbins' compound decision procedure. *AMS*, **43**, 348–350.

Huber, P.J. (1964). Robust estimation of a location parameter. *AMS*, **35**, 73–101.

Ipsen, J. and Jerne, N.K. (1944). Graphical evaluation of the distribution of small experimental series. *Acta Pathologica et Microbiologica Scandinavica*, **21**, 343–361.

Ivory, J. (1825). On the method of the least squares. *Phil. Mag.*, **65**, 3–10.

Jaeckel, L.A. (1971). Robust estimates of location: Symmetry and asymmetric contamination. *AMS*, **42**, 1020–1034.

Jaynes, E.T. (1957). Information theory and statistical mechanics. *Phys. Rev.*, **106**, 620–630.

Jeffreys, H. (1939). *Theory of Probability*. Oxford University Press.

Jevons, W.S. (1875). *Money and the Mechanism of Exchange*. International Scientific Series, London and New York.

Jorland, G. (1987). The Saint Petersburg Paradox 1713–1937. In L. Krüger, L.J. Daston, and M. Heidelberger (eds.), *The Probabilistic Revolution*, Vol. 1. MIT Press, Cambridge, MA, pp. 157–190.

Kac, M. (1945). Random walk in the presence of absorbing barriers. *AMS*, **16**, 62–67.

Kalbfleisch, J.D. (1978). Non-parametric Bayesian analysis of survival time data. *JRSS B*, **40**, 214–221.

Kalbfleisch, J.D. and Prentice, R.L. (1973). Marginal likelihoods based on Cox's regression and life model. *Biometrika*, **60**, 267–278.

Kaplan, E.L. and Meier, P. (1958). Nonparametric estimation from incomplete observations. *JASA*, **53**, 457–481.

Kapteyn, J.C. (1903). *Skew Frequency Curves in Biology and Statistics*. Groningen.

Kendall, D.G. (1948). On some modes of population growth leading to R.A. Fisher's logarithmic series distribution. *Biometrika*, **35**, 6–15.

Kendall, M.G. (1940). Note on the distribution of quantiles for large samples. *Suppl. JRSS*, **7**, 83–85.

Kendall, M.G. and Plackett, R.L. (eds.) (1977). *Studies in the History of Statistics and Probability*, II. Griffin, London.

Keynes, J.M. (1921). *A Treatise on Probability*. Macmillan, London.

Khintchine, A. (1934). Korrelationstheorie der stationären stochastischen Prozesse. *Math. Ann.*, **109**, 604–615.

King, W.I. (1912). *The Elements of Statistical Method*. Macmillan, New York.

Knies, C.G.A. (1850). *Die Statistik als selbstständige Wissenschaft*. Luckhardt, Kassel.

Kolmogoroff, A. (1929). Das Gesetz des iterierten Logarithmus. *Math. Ann.*, **101**, 126–135.

Kolmogorov, A. (1933). Sulla determinazione empirica di una legge de distribuzione. *Giornale dell' istituto italiano degli attuari*, **4**, 83–91.

Kolmogorov, A.N. and Dmitriev, N.A. (1947). Branching stochastic processes. Doklady Akademii Nauk, USSR, **56**, 5–8.

Kolodziejczyk, S. (1935). On an important class of statistical hypotheses. *Biometrika*, **27**, 161–190.

Koopman, B.O. (1936). On distributions admitting a sufficient statistic. *Trans. Amer. Math. Soc.*, **39**, 399-409.

Koopmans, T.C. (1949). Identification problems in economic model construction. *Econometrica*, **17**, 125–144.

Kotz, S., Johnson, N.L., and Read, C. (eds.) (1982–1989). *Encyclopedia of Statistical Sciences*. Wiley, New York.

Kruskal, W.H. and Stigler, S.M. (1997). Normative terminology: 'normal' in statistics and elsewhere. In B.D. Spencer (ed.), *Statistics and Public Policy*. Clarendon Press, Oxford.

Kruskal, W.H. and Tanur, J.M. (eds.) (1978). *International Encyclopedia of Statistics*. Free Press, New York.

Kullback, S. (1934). An application of characteristic functions to the distribution problem of statistics. *AMS*, **5**, 263–307.

Kurtz, A.K. and Edgerton, H.A. (1939). *Statistical Dictionary of Terms and Symbols*. Wiley, New York.

Lancaster, H.O. (1949). The derivation and partition of χ^2 in certain discrete distributions. *Biometrika*, **36**, 117–129.

LeCam, L. (1953). On some asymptotic properties of maximum likelihood estimates and related Bayes' estimates. *University of California, Publications in Statistics*, **1**, 277-330.

Legendre, A.M. (1805). *Nouvelles méthodes pour la détermination des orbites des comètes*. F. Didot, Paris.

Lehmann, E.L. (1953). The power of rank tests. *AMS*, **24**, 23–43.

Lehmann, E.L. (1983). *Theory of Point Estimation*. Wiley, New York.

Lehmann, E.L. and Scheffé, H. (1950). Completeness, similar regions, and unbiased estimation. *Sankhyā*, **10**, 305–340.

Lehmann, E.L. and Stein, C. (1948). Most powerful tests of composite hypotheses. I. Normal distributions. *AMS*, **19**, 495–516.

Lehmann, E.L. and Stein, C. (1950). Completeness in the sequential case. *AMS*, **21**, 376–385.

Lévy, P. (1923). Sur les lois stables en calcul des probabilités. *CRAS*, **176**, 1284–1286.

Lévy, P. (1925). *Calcul des Probabilités*. Gauthier-Villars, Paris.

Lieblein, J. (1953). On the exact evaluation of the variances and covariances of order statistics in samples from the extreme-value distribution. *AMS*, **24**, 282–287.

Lieblein, J. (1955). On moments of order statistics from the Weibull distribution. *AMS*, **26**, 330–333.

Lloyd, H. (1848). On certain questions connected with the reduction of magnetical and meteorological observations. *Proc. Roy. Irish Acad.*, **4**, 180–183.

Loève, M. (1955). *Probability Theory*. Van Nostrand, Princeton, NJ.

Lubbock, J.W. and Drinkwater-Bethune, J.E. (c. 1830). *On Probability*. Library of Useful Knowledge. Baldwin and Cradock, London.

Mann, H.B. and Whitney, D.R. (1947). On a test of whether one of two random variables is stochastically larger than the other. *AMS*, **18**, 50–60.

Markoff, A.A. (1912). *Wahrscheinlichkeitsrechnung*. Teubner, Leipzig.

Marriott, F.H.C. (1990). *A Dictionary of Statistical Terms*, 5th edn. Longman, London.

Marshall, A.W. and Proschan, F. (1972). Classes of distributions applicable in replacement, with renewal theory implications. *PBS6*, Vol. 1, pp. 395–415.

Matheron, G. (1963a). *Traité de géostatistique appliquée*, Tome 2: *Le krigeage*. Bureau de recherche géologiques et minières, Paris.

Matheron, G. (1963b). Principles of geostatistics. *Economic Geology*, **58**, 1246–1266.

McAlister, D. (1879). The law of the geometric mean. *PRSL*, **29**, 367–376.

McCulloch, J.H. (1985). On heteros* edasticity. *Econometrica*, **53**, 483.

McIntyre, G.A. (1952). A method for unbiased selective sampling using ranked sets. *Austr. J. Agric. Res.*, **3**, 385–390.

Merriman, M. (1877–1882). A list of writings relating to the method of least squares, with historical and critical notes. *Trans. Connecticut Acad. Arts and Sciences*, **4**, 151–231. [Reproduced by the Statistical Engineering Laboratory of the National Bureau of Standards, Mar. 1957.]

Metropolis, N. and Ulam, S. (1949). The Monte Carlo method. *JASA*, **44**, 335–341.

Miller, L.H. (1956). Table of percentage points of Kolmogorov statistics. *JASA*, **51**, 111–121.

Miller, R.G. (1964). A trustworthy jackknife. *AMS*, **35**, 1594–1605.

Mises, R. von (1919). Grundlagen der Wahrscheinlichkeitsrechnung. *Math. Zeit.*, **5**, 52–99.

Montricher, G.F. de, Tapia, R.A., and Thompson, J.R. (1975). Nonparametric maximum likelihood estimation of probability densities by penalty function methods. *AS*, **3**, 1329–1348.

Mood, A.M. (1950). *Introduction to the Theory of Statistics*. McGraw-Hill, New York.

Mosteller, F. (1946). On some useful "inefficient" statistics. *AMS*, **17**, 377–408.

Mosteller, F. (1948). A k-sample slippage test for an extreme population. *AMS*, **19**, 58–65.

Nelder, J.A. (1974). Log linear models for contingency tables: A generalization of classical least squares. *Appl. Statist.*, **23**, 323–329.

Nelder, J.A. and Wedderburn, R.W.M. (1972). Generalized linear models. *JRSS* A, **135**, 370–384.

Neumann, J. von (1928). Zur Theorie der Gesellschaftsspiele. *Math. Annalen*, **100**, 295–320.

Neumann, J. von (1941). Distribution of the ratio of the mean square successive difference to the variance. *AMS*, **12**, 367–395.

Neumann, J. von and Morgenstern, O. (1944). *Theory of Games and Economic Behavior*. Princeton University Press.

Neyman, J. (1934). On the two different aspects of the representative method. *JRSS*, **97**, 558-625.

Neyman, J. (1939). On a new class of "contagious" distributions, application in entomology and bacteriology. *AMS*, **10**, 35–57.

Neyman, J. (1950). *First Course in Probability and Statistics*. Holt, New York.

Neyman, J. and Pearson, E.S. (1928). On the use and interpretation of certain test criteria for purposes of statistical inference, Part I. *Biometrika*, **20** A, 175–240.

Neyman, J. and Pearson, E.S. (1931). On the problem of k samples. *Bull. acad. Polonaise des sciences et lettres*, Ser. A, 460–481.

Neyman, J. and Pearson, E.S. (1933a). On the problem of the most efficient tests of statistical hypotheses. *PTRSL* A, **231**, 289–337.

Neyman, J. and Pearson, E.S. (1933b). The testing of statistical hypotheses in relation to probabilities *a priori*. *PCPS*, **24**, 492–510.

Neyman, J. and Pearson, E.S. (1936). Contributions to the theory of testing statistical hypotheses. *SRM*, **1**, 1–37.

Neyman, J. and Scott, E.L. (1948). Consistent estimates based on partially consistent observations. *Econometrica*, **16**, 1–32.

Nishisato, S. (1980). *Analysis of Categorical Data: Dual Scaling and Its Applications*. University of Toronto Press.

Noether, G.E. (1950). Asymptotic properties of the Wald–Wolfowitz test of randomness. *AMS*, **21**, 231–246.

Page, E.S. (1954). Continuous inspection schemes. *Biometrika*, **41**, 100–115.

Parzen, E. (1960). *Modern Probability Theory and Its Applications*. Wiley, New York.

Parzen, E. (1979). Nonparametric statistical data modeling. *JASA*, **74**, 105–121.

Patterson, H.D. and Williams, E.R. (1976). A new class of resolvable incomplete block designs. *Biometrika*, **63**, 83–92.

Pearson, E.S. (1939). William Sealy Gosset, 1876–1937. "Student" as statistician. *Biometrika*, **30**, 210–250.

Pearson, E.S. (1941). Tables of percentage-points of the incomplete beta-function. Prefatory note. *Biometrika*, **32**, 151–153.

Pearson, E.S. and Hartley, H.O. (1943). Tables of the probability integral of the studentized range. *Biometrika*, **33**, 89–99.

Pearson, E.S. and Kendall, M.G. (eds.) (1970). *Studies in the History of Statistics and Probability*. Hafner, Darien, CT.

Pearson, K. (1893). *Nature*. Oct. 26, 615.

Pearson, K. (1894). Contributions to the mathematical theory of evolution. *PTRSL* A, **185**, 71–110.

Pearson, K. (1895). Skew variation in homogeneous material. *PTRSL* A, **186**, 343–414.

Pearson, K. (1896). Regression, heredity and panmixia. *PTRSL* A, **187**, 253–318.

Pearson, K. (1900). On the criterion that a given system of deviations from the probable in the case of a correlated system of variables is such that it can be reasonably supposed to have arisen from random sampling. *Phil. Mag.*, 5th Ser., **50**, 157–175.

Pearson, K. (1901). Supplement to a memoir on skew variation. *PTRSL* A, **197**, 443–459.

Pearson, K. (1902). On the systematic fitting of curves to observations and measurements. *Biometrika*, **1**, 265–303.

Pearson, K. (1904). On the theory of contingency and its relation to association and normal correlation. *Drapers' Company Research Memoirs*, Biometric Ser. I.

Pearson, K. (1905a). On the general theory of skew correlation and nonlinear regression. *Drapers' Company Research Memoirs*, Biometric Ser. II.

Pearson, K. (1905b). Das Fehlergesetz und seine Verallgemeinerungen durch Fechner und Pearson. A rejoinder. *Biometrika*, **4**, 169–212.

Pearson, K. (1905c). The problem of the random walk. *Nature*, July 27, 294.

Pearson, K. (1908). On the generalised probable error in multiple normal correlation. *Biometrika*, **6**, 59–68.

Pearson, K. (1909). On a new method of determining correlation... *Biometrika*, **7**, 96–105.

Pearson, K. (1914). On certain errors with regard to multiple correlation occasionally made by those who have not adequately studied this subject. *Biometrika*, **10**, 181–187.

Pearson, K. (1920). Notes on the history of correlation. *Biometrika*, **13**, 25–45.

Pearson, K. (1927). Foreword to *Random Sampling Numbers*, arranged by L.H.C. Tippett. *Tracts for Computers XV*. University College, London.

Pearson, K. (1956). *Karl Pearson's Early Statistical Papers*. Cambridge University Press.

Pearson, K. and Lee, A. (1897). On the distribution of frequency (variation and correlation) of the barometric height of divers stations. *PTRS*, **190**, 423–469.

Pearson, K. and Lee, A. (1908). Generalized probable error in multiple normal correlation. *Biometrika*, **6**, 59–68.

Peirce, C.S. (1873). On the theory of errors of observations. Appendix No. 21 of the Report of the Superintendent of the U.S. Coast Survey for the Year Ending June 1870, pp. 200-224. [Reprinted in Stigler (1980b).]

Persons, W.M. (1919). An index of general business conditions. *Rev. Econ. Statist.*, **1**, 113–151.

Pigou, A.C. (1920). *Economics of Welfare*. Macmillan, London.

Pitman, E.J.G. (1937). Significance tests which may be applied to samples from any populations. III. The analysis of variance test. *Biometrika*, **29**, 322–335.

Pitman, E.J.G. (1938). The estimation of the location and scale parameters of a continuous population of any given form. *Biometrika*, **30**, 391–421.

Plackett, R.L. (1972). Studies in the history of probability and statistics. XXIX. The discovery of the method of least squares. *Biometrika*, **59**, 239–251.

Poincaré, H. (1912). *Calcul des probabilités*. Gauthier-Villars, Paris.

Poisson, S.D. (1835). Recherches sur la probabilité des jugements, principalement en matiére criminelle. *CRAS*, **1**, 473–494.

Pollack, I. and Decker, L.R. (1958). Confidence ratings, message reception, and the receiver operating characteristic. *J. Acoust. Soc. Amer.*, **30**, 286–292.

Pólya, G. (1920). Über den zentralen Grenzwertsatz der Wahrscheinlichkeitsrechnung und das Momentenproblem. *Math. Zeit.*, **8**, 171–181.

Pritzker, L., Ogus, J., and Hansen, M.H. (1965). Computer editing methods — some applications and results. *Bull. Intern. Statist. Inst.*, **41**, 442–466.

Raiffa, H. and Schlaifer, R. (1961). *Applied Statistical Decision Theory*. Harvard Business School, Cambridge, MA.

Ramsay, W. (1892). Report of a paper read to the Chemical Society, London. *Nature*, **45**, 429/2.

Rao, C.R. (1948). Large sample tests of statistical hypotheses concerning several parameters with applications to problems of estimation. *PCPS*, **44**, 50–57.

Rao, C.R. (1957). Theory of the method of estimation by minimum chi-square. *Bull. Inst. Intern. Statist.*, **35**, 2, 25–32.

Robbins, H. (1951). Asymptotically subminimax solutions of compound statistical decision problems. *PBS2*, 131–148.

Robbins, H. (1956). An empirical Bayes approach to statistics. *PBS3*, Vol. 1, 157–163.

Rubin, H. (1951). A complete class of decision procedures for distributions with monotone likelihood ratio (Abstract). *AMS*, **22**, 608.

Ryan, T.A., Jr. (1975). Robust regression — bounded leverage. In *Proc. Statist. Comp. Section*. Amer. Statist. Assn., pp. 138–141.

Savage, L.J. (1954). *The Foundations of Statistics*. Wiley, New York.

Scheffé, H. (1956). Alternative models for the analysis of variance. *AMS*, **27**, 251–271.

Scheffé, H. (1959). *The Analysis of Variance*. Wiley, New York.

Schuster, A. (1898). On the investigation of hidden periodicities with application to a supposed 26 day period of meteorological phenomena. *Terrestrial Magnetism*, **3**, 13–41.

Seal, H.L. (1949). The historical development of the use of generating functions in probability theory. *Bull. ass. des actuaires Suisses*, **49**, 209–228. [Reprinted in Kendall and Plackett (1977).]

Seal, H.L. (1967). The historical development of the Gauss linear model. *Biometrika*, **54**, 1–24.

Shewhart, W.A. (1931). *Economic Control of Quality of Manufactured Product*. Van Nostrand, New York.

Sheynin, O. (1997). Letter to *The American Statistician*, **51**, 210.

Simaika, J.B. (1941). On an optimum property of two important statistical tests. *Biometrika*, **32**, 70 80.

Smirnoff, N. (1933). Ueber Wahrscheinlichkeiten grosser Abweichungen. *Recueil Soc. Math. Moscou*, **40**, 441–454.

Snedecor, G.W. (1934). *Calculation and Interpretation of Analysis of Variance and Covariance*. Collegiate Press, Ames, IA.

Soper, H.E. (1914). Tables of Poisson's exponential binomial limit. *Biometrika*, **10**, 25–35.

Spearman, C. (1914). The theory of two factors. *Psychol. Rev.*, **21**, 101–115.

Steffensen, J.F. (1923). Factorial moments and discontinuous frequency-functions. *SA*, **6**, 73–89.

Stewart, W.M. (1941). A note on the power of the sign test. *AMS*, **12**, 124.

Stigler, S.M. (ed.) (1980b). *American Contributions to Mathematical Statistics in the Nineteenth Century*, Vol. II. Arno, New York.

Stigler, S.M. (1986). *The History of Statistics*. Harvard University Press, Cambridge, MA.

Stuart, A. and Ord, K.J. (1987). *Kendall's Advanced Theory of Statistics*, Vol. 1, 5th edn. Oxford University Press.

Student (1908). The probable error of a mean. *Biometrika*, **6**, 1–25.

Sylvester, J.J. (1851). On a remarkable discovery in the theory of canonical forms and of hyperdeterminants. *Phil. Mag.*, 4th Ser., **2**, 391–410.

Tang, P.C. (1938). The power function of the analysis of variance tests with tables and illustrations of their use. *SRM*, **2**, 126–157.

Thompson, J.R. (1968). Some shrinkage techniques for estimating the mean. *JASA*, **63**, 113–122.

Thurstone, L.L. (1927). The method of paired comparisons for social values. *J. Abnormal and Social Psychol.*, **21**, 384–400.

Thurstone, L.L. (1931). Multiple factor analysis. *Psychol. Rev.*, **38**, 406–427.

Todhunter, I. (1865). *A History of the Mathematical Theory of Probability.* Macmillan, London. [Reprinted 1949, 1965, Chelsea, New York.]

Torgerson, W.S. (1952). Multidimensional scaling: I. Theory and method. *Psychometrika*, **17**, 401–419.

Tryon, R.C. (1939). *Cluster Analysis.* Edwards Brothers, Ann Arbor, MI.

Tukey, J.W. (1949). The simplest signed-rank tests. Statistical Research Group Mimeo Rep. No. 17. Princeton University.

Tukey, J.W. (1962). The future of data analysis. *AMS*, **33**, 1–67.

Tukey, J.W. (1970). *Exploratory Data Analysis*, Vol. 1. Addison-Wesley, Reading, MA (limited preliminary edition).

Tukey, J.W. (1972). Some graphic and semigraphic displays. In *Statistical Papers in Honor of George W. Snedecor*, T.A. Bancroft (ed.), Iowa State University Press, Ames, IA, pp. 293–316.

Tweedie, M.C.K. (1945). Inverse statistical variates. *Nature*, **155**, 453.

Tweedie, M.C.K. (1947). Functions of a statistical variate with given means, with special reference to Laplacian distributions. *PCPS*, **43**, 41–49.

Uspensky, J.V. (1937). *Introduction to Mathematical Probability.* McGraw-Hill, New York.

Valavanis, S. (1959). *Econometrics.* McGraw-Hill, New York.

Vautel, J.W., Manton, K., and Stallard, E. (1979). The impact of hetero-geneity in individual frailty on the dynamics of mortality. *Demography*, **16**, 439–454.

Verhulst, P.-F. (1845). La loi d'accroissement de la population. *Nouveaux mémoires de l'académie royale des sciences et belles-lettres de Bruxelles*, **18**, 1–38.

Ville, J. (1939). *Étude critique de la notion de collectif.* Gauthier-Villars, Paris.

Wald, A. (1939). Contributions to the theory of statistical estimation and testing hypotheses. *AMS*, **10**, 299–326.

Wald, A. (1941). Asymptotically most powerful tests of statistical hypothe-ses. *AMS*, **12**, 1–19.

Wald, A. (1943). Sequential analysis of statistical data: Theory. Report to the Applied Mathematics Panel, National Defense Research Committee.

Wald, A. (1945a). Statistical decision functions which minimize the maxi-mum risk. *AMS*, **46**, 265–280.

Wald, A. (1945b). Sequential tests of statistical hypotheses. *AMS*, **16**, 117–186.

Wald, A. (1947a). An essentially complete class of admissible decision functions. *AMS*, **18**, 549–555.

Wald, A. (1947b). Foundations of a general theory of sequential decision functions. *Econometrica*, **15**, 279–313.

Wald, A. (1947c). *Sequential Analysis*. Wiley, New York.

Wald, A. (1948). Asymptotic properties of the maximum likelihood estimate of an unknown parameter of a discrete stochastic process. *AMS*, **19**, 40–46.

Wald, A. (1950). *Statistical Decision Functions*. Wiley, New York.

Wald, A. and Wolfowitz, J. (1940). On a test whether two samples are from the same population. *AMS*, **11**, 147–162.

Walker, H. (1929). *Studies in the History of Statistical Method*. Williams and Wilkins, Baltimore.

Warner, S.L. (1965). Randomized response: A survey technique for eliminating evasive answer bias. *JASA*, **60**, 63–69.

Weatherburn, C.E. (1946). *A First Course in Mathematical Statistics*. Cambridge University Press.

Wedderburn, R.W.M. (1974). Quasi-likelihood functions, generalized linear models, and the Gauss–Newton method. *Biometrika*, **61**, 439–447.

Wegman, E.J. (1972). Nonparametric probability density estimation: I. A summary of available methods. *Technometrics*, **14**, 533–546.

West, C.J. (1918). *Introduction to Mathematical Statistics*. R.G. Adams, Columbus, OH.

Westergaard, H. (1932). *Contributions to the History of Statistics*. King, London.

Whewell, W. (1831). In *Letters II (1876)*. Macmillan, London, p. 135 (Nov. 12).

Wijsman, R.A. (1967). Cross-sections of orbits and their application to densities of maximal invariants. *PBS5*, Vol. 1, 389–400.

Wilcoxon, F. (1945). Individual comparisons by ranking methods. *Biometrics Bull.*, **1**, 80–83.

Wilk, M.B. and Gnanadesikan, R. (1968). Probability plotting methods for the analysis of data. *Biometrika*, **55**, 1–17.

Wilks, S.S. (1941). Determination of sample sizes for setting tolerance limits. *AMS*, **12**, 91–96.

Wilks, S.S. (1942). Statistical prediction with special reference to the problem of tolerance limits. *AMS*, **13**, 400–409.

Wilks, S.S. (1943). *Mathematical Statistics*. Princeton University Press.

Williams, C.B. (1944). Some applications of the logarithmic series and the index of diversity to ecological problems. *J. Ecology*, **32**, 1–44.

Williams, J.D. (1954). *The Compleat Strategyst*. McGraw-Hill, New York.

Williams, S.R. (1903). Variation in lithobius forficatus. *Amer. Naturalist*, **37**, 299–312.

Winter, A. (1934). On analytic convolutions of Bernoulli distributions. *Amer. J. Math.*, **56**, 659–663.

Wishart, J. (1928). The generalised product moment distribution in samples from a normal multivariate population. *Biometrika*, **20A**, 32–52.

Wold, H. (1938). *A Study in the Analysis of Stationary Time Series*. Almqvist and Wiksell, Stockholm.

Wolfowitz, J. (1942). Additive partition functions and a class of statistical hypotheses. *AMS*, **13**, 247–279.

Wood, H.C. (1912). The purpose and limitations of bio-assay. *J. Amer. Medical Assn.*, **59**, 1433–1434.

Wright, S. (1921). Correlation and causation. *J. Agric. Res.* (Washington, DC), **20**, 557–585.

Wrinch, D. and Jeffreys, H. (1921). On certain fundamental principles of scientific inquiry. *Phil. Mag.*, **42**, 6th Ser., 369–390.

Wurtele, Z.S. (1949). Continuous sampling plans from the risk point of view (Abstract). *AMS*, **20**, 469.

Yates, F. (1933). The principles of orthogonality and confounding in repliated experiments. *J. Agric. Sci.*, **23**, 108–145.

Yates, F. (1935). Complex experiments. *JRSS*, Suppl. 2, 181–233.

Yates, F. (1937). *The Design and Analysis of Factorial Experiments*. Imperial Bureau of Soil Science, Techn. Comm. No. 35, Harpenden.

Yates, F. (1939). The recovery of inter-block information in variety trials arranged in three-dimensional lattices. *AE*, **9**, 136–156.

Yule, G.U. (1897). On the theory of correlation. *JRSS*, **60**, 812–854.

Yule, G.U. (1900). On the association of attributes in statistics. *PTRSL* A, **194**, 257–319.

Yule, G.U. (1902). Mendel's laws and their probable relations to intraracial heredity. *New Phytologist*, **1**, 193–207.

Yule, G.U. (1907). On the theory of correlation for any number of variables treated by a new system of notation. *PRS* A, **79**, 182–193.

Yule, G.U. (1911). *An Introduction to the Theory of Statistics*. Griffin, London.

Yule, G.U. (1926). Why do we sometimes get nonsense-correlations between time-series? *JRSS*, **89**, 1–64.

Note: My colleague W.Q. Meeker has recently drawn to my attention the possibility of searching for terms in the electronic journal archive JSTOR. At present JSTOR's coverage is confined to a limited number of leading journals in various fields including mathematics and statistics. For statistics and probability JSTOR can not as yet match the much wider search represented by our list. However, some sampling clearly shows that JSTOR would help improve quite a few of the entries, usually by just a few years. Before accepting any of the attributions on our list as the best available knowledge, the reader is advised to check against the latest version of JSTOR.

Subject Index

The appendices are not indexed.
Bold type refers to the extracts of original material.

Abbe's Criterion, 77
accuracy of observations, 37–40, **41**
aleatory contracts, 2
arrangements, 20, **26–29**
Ars conjectandi, 3
association of attributes, 137, **141**,
      measure of, 137, partial, 137,
      illusory, 138 **142**
asymptotic relative efficiency, 51–55

Bayesian confidence interval, 38
Bayesian methods, 38, 131–132, 187–
      190, **194–201**
Bayes's rule, 131, **134**
Bernoulli's theorem. *See* law of large
      numbers
Bernoulli trials, 146. *See also* bino-
      mial distribution
binomial coefficients, **13**, **14**, 132, 133
binomial distribution, 9, 10, **13**, **14**,
      115, 130–132, **134**, contin-
      uous, 132–133
binomial parameter, estimation of,
      129–133, **134–136**, 187, 191
births, male and female, 9, **13–17**
blind fate, 8, 11
Boyle Lectures, 7
Bradley-Terry model. *See* paired-
      comparison model

central limit theorem, 38, 51
Chance and Design, 7
chi-squared distribution, 103–107
circular serial correlation coefficient,
      77
coincidences, 19–23
combinations, 21
comets, 8
confidence coefficient, 191
confidence concept, 187, 190

confidence distribution, 191
confidence interval, 38, 187–193
confidence limits. *See* confidence in-
      terval
consistency, 130
contingency table, 137, 138
contour integration, 78
correlation, spurious, 137, **143**
corelation coefficient, 137, 188, **200**,
      201

De ratiociniis in ludo aleæ, 2, 9
design, argument from, 7, **118–121**,
      129. *See also* Divine Prov-
      idence
direct probability methods, 38, 116,
      131, 187–189, **196**
Divine Providence, **13**
division problem. *See* Problem of Points
drawn games. *See* ties

Empirical Logic, 116
equations of condition, 51, **57 59**
Essay on the Usefulness of Mathe-
      matical Learning, 9
Essay on the Principle of Popula-
      tion, 65
Euler diagram, 115
Evidences, 7
exchangeable events, 19, 21, 22
expectation, 1–6, **15**, **26–28**, **135**,
      151
expected value. *See* expectation
exponential growth, 65, 66, **71**, **74**
extreme-value theory, 145–150

fair game, 1
fiducial distribution, 188, 189, **200**,
      201
fiducial limits, 191

fiducial probability, 188–192, **199–201**

fractal, 115–116

gamble, value of, 1, 2
Game of Thirteen, 20, **25**, **31**
Gaussian law of error, **109**, **151**, **155**, **159**
goodness-of-fit statistics, 77–79. *See also* chi-squared distribution

half-invariants, 130
Helmert transformation, 103, 104
homogeneity of variance, test of, 22
hypothesis-testing. *See* Neyman–Pearson theory

inclusion and exclusion, method of, 19–23
insufficient reason, doctrine of, **194**
interval estimation. *See* confidence interval; fiducial limits
inverse probability, 37, 38, **119**, 188, 189, **194**

L_1- and L_2-estimation, 51
law of large numbers, **127**, 147, **158**
least squares, 37, **41**, 51, 53, **82**, 105
Lectures on the General Theory of Observations, 129, **134**
likelihood, 129–133, **134**, 187, **198**. *See also* maximum likelihood
likelihood function, 129–133
likelihood ratio, 10
linear model, 51
linear regression, 51, 52
logistic distribution, 66
logistic growth curve, 65–67, **73**, **74**

males, excess of, 9–11, **15**
matches. *See* coincidences
maximum likelihood, 38, 129, 130, 132, 161, 162, 188, **198**
maximum probability, 38, **41**, **43**, **46**, **48**, 188, **197**
Mécanique céleste, **57**
median, 51–53, **196**, **197**

median absolute deviation, 38
mean deviation, 103
mean difference, 103
mean error, **86**, 103, **109**, **112**, **113**
mean squared error, **112**
mode, 130, **134**, 188, **196**, **197**
most advantageous method. *See* least squares

Neyman-Pearson theory, 191
null hypothesis, 7, 11

odds ratio, 138
Of the Laws of Chance, 9
Opticks, 8
order statistics, 51, 52, 146

paired-comparison model, 161, 164; method, 162
parameter, **195–199**
Pascal's wager, 3
pi (π), randomness of digits, **121–128**
planets, 8, 10
population growth, law of, **69**
precision, 37, **43**, **82**, 103
probable error, 42, **46**, **82**, 103, **109**, 187
probable limits, **44–48**
Problem of Points, 1–3, **4**

Q, 139

randomness, **117**
random walk, 115–116; figure, **118**
range, 145, 146, **151**, **155**
recursion, 1–3
regression coefficient, 51
regular tournaments. *See* Round Robins
rejection region, 11
repeated-sampling methods. *See* direct probability methods
Round Robins, 161–165
Royal Society, 9
Royal Society theology, 7, 10

sample variance, distribution of, 103–107
significance level, 188

Simpson's paradox. *See* Yule's para-
    dox
situation, method of, 51, **61**, **63**, **64**
stake, 1–3
standard deviation, 105, **151**, **155**
stars, arrangement of, **118**, **119**
statistic, **198–200**
Stigler's Law, 138
sum of squares, **82**, **87**, **98**, **111**
support, 129
Symbolic Logic, 116

t-distribution, 187
test of significance, 7–12, 188
The Doctrine of Chances, 7, 10, 21,
    32
The Logic of Chance, 115, **117**
*The Genetical Theory of Natural Se-
    lection*, 11

Theoria Motus Corporum Coelestium,
    41, **43**
Théorie analytique des probabilités,
    51, 53, **61–63**
ties, 164–165
tournament outcomes, 161–166, **167**
Traité du triangle arithmétique, 2, 3,
    5

variance, estimation of, 105
Venn diagram, 115

Wahrscheinlichkeit, Statistik und Wahrheit,
    148
weighted median estimator, 51–53

Yule's coefficient *Q*, 139
Yule's paradox, 137–140

Name Index

Commentators, modern contributors, and names in the appendices are not indexed. Bold type refers to the extracts of original material.

Abbe, 77–79, **81–101**, 104
Anne, Queen, 9
Arbuthnott, 7–12, **13–17**

Bayes, 37, 131, **134, 135**, 189, **194**
Bentley, 7, 8, 10
Bernoulli, Daniel, 10, 129
Bernoulli, James, 3, 10, **127, 158**
Bernoulli, Nicholas, 10, 19–23, **31**, 145
Bessel, 37, **48, 49**
Bienaymé, 77, 104
Bing, 131, **135**
Boole, **197**
Bortkiewicz, 145–150, **151–153, 155, 156, 159**
Boscovich, 52
Bull, 9

Carcavi, 2, **4**
Cauchy, 78, **84, 89**
Clarke, 7
Clopper, 191
Cochran, 22
Cournot, 187

Darwin, 7, 11
De Méré, **4**
De Moivre, 7, 9, 10, 19–23, **32–36**
Derham, 7, 9, 10
Dirichlet, 38, 78, **84**
Dodd, 147

Encke, 38
Ezekiel, **200**

Fermat, 1–4
Fisher, 7, 11, 51, 104, 105, 116, 129, 130, 147, 148, 187–193, **194–201**

Fréchet, 147
Frobenius, 164

Gauss, 37–40, **41–49**, 77, **81**, 103–105, 145, 187, **194, 197**
Gnedenko, 147, 148
Gumbel, 147

Hauber, 38
Helmert, 77, 103–107, **109–113**
Hitler, 149, 165
Huygens, 2, 3

Jacobi, **89**

Lambert, 129
Landau, 163, 164
Laplace, 37, 38, **45**, 51–55, **57–64**, 134, 187, **194**
Legendre, 52

Malthus, 65, **72**
Montmort, 10, 19–23, **25–29**, 31

Newton, 7–9, **125**
Neyman, 148, 187–192

Paley, 7
Pascal, 1–4, **4**, 115
Pearson, E.S., 187, 191
Pearson, Karl, 104, 105, 115, 116, 137, **143**, 146
Pizzetti, 104, 187
Planck, 165
Poisson, 66, 78
Price, **194**

Quetelet, 65

Rayleigh, 116
Riemann, **85**
Rilke, 148
Roberval, 2, **4**

'sGravesande, 10
"Student," 78, 187, 188
Swift, 9

Tartaglia, 1
Thiele, 129–133, **134–136**
Thurstone, 162

Tippett, 147, 148

Venn, 115–116, **117–128**, **197**
Verhulst, 65–67, **69–75**
von Mises, **155–160**

Yule, 137–140, **141–143**

Zeiss, 78
Zermelo, 161–166, **167–186**

Springer Series in Statistics

Andersen/Borgan/Gill/Keiding: Statistical Models Based on Counting Processes.
Atkinson/Riani: Robust Diagnotstic Regression Analysis.
Berger: Statistical Decision Theory and Bayesian Analysis, 2nd edition.
Bolfarine/Zacks: Prediction Theory for Finite Populations.
Borg/Groenen: Modern Multidimensional Scaling: Theory and Applications
Brockwell/Davis: Time Series: Theory and Methods, 2nd edition.
Chen/Shao/Ibrahim: Monte Carlo Methods in Bayesian Computation.
David/Edwards: Annotated Readings in the History of Statistics.
Efromovich: Nonparametric Curve Estimation: Methods, Theory, and Applications.
Fahrmeir/Tutz: Multivariate Statistical Modelling Based on Generalized Linear
    Models.
Farebrother: Fitting Linear Relationships: A History of the Calculus of Observations
    1750-1900.
Federer: Statistical Design and Analysis for Intercropping Experiments, Volume I:
    Two Crops.
Federer: Statistical Design and Analysis for Intercropping Experiments, Volume II:
    Three or More Crops.
Fienberg/Hoaglin/Kruskal/Tanur (Eds.): A Statistical Model: Frederick Mosteller's
    Contributions to Statistics, Science and Public Policy.
Fisher/Sen: The Collected Works of Wassily Hoeffding.
Good: Permutation Tests: A Practical Guide to Resampling Methods for Testing
    Hypotheses, 2nd edition.
Gouriéroux: ARCH Models and Financial Applications.
Grandell: Aspects of Risk Theory.
Haberman: Advanced Statistics, Volume I: Description of Populations.
Hall: The Bootstrap and Edgeworth Expansion.
Härdle: Smoothing Techniques: With Implementation in S.
Hart: Nonparametric Smoothing and Lack-of-Fit Tests.
Hartigan: Bayes Theory.
Hedayat/Sloane/Stufken: Orthogonal Arrays: Theory and Applications.
Heyde: Quasi-Likelihood and its Application: A General Approach to Optimal
    Parameter Estimation.
Huet/Bouvier/Gruet/Jolivet: Statistical Tools for Nonlinear Regression: A Practical
    Guide with S-PLUS Examples.
Kolen/Brennan: Test Equating: Methods and Practices.
Kotz/Johnson (Eds.): Breakthroughs in Statistics Volume I.
Kotz/Johnson (Eds.): Breakthroughs in Statistics Volume II.
Kotz/Johnson (Eds.): Breakthroughs in Statistics Volume III.
Küchler/Sørensen: Exponential Families of Stochastic Processes.
Le Cam: Asymptotic Methods in Statistical Decision Theory.
Le Cam/Yang: Asymptotics in Statistics: Some Basic Concepts, 2nd edition.
Longford: Models for Uncertainty in Educational Testing.
Miller, Jr.: Simultaneous Statistical Inference, 2nd edition.
Mosteller/Wallace: Applied Bayesian and Classical Inference: The Case of the
    Federalist Papers.
Parzen/Tanabe/Kitagawa: Selected Papers of Hirotugu Akaike.

Springer Series in Statistics

Politis/Romano/Wolf: Subsampling.
Ramsay/Silverman: Functional Data Analysis.
Rao/Toutenburg: Linear Models: Least Squares and Alternatives.
Read/Cressie: Goodness-of-Fit Statistics for Discrete Multivariate Data.
Reinsel: Elements of Multivariate Time Series Analysis, 2nd edition.
Reiss: A Course on Point Processes.
Reiss: Approximate Distributions of Order Statistics: With Applications
    to Non-parametric Statistics.
Rieder: Robust Asymptotic Statistics.
Rosenbaum: Observational Studies.
Rosenblatt: Gaussian and Non-Gaussian Linear Time Series and Random Fields.
Särndal/Swensson/Wretman: Model Assisted Survey Sampling.
Schervish: Theory of Statistics.
Shao/Tu: The Jackknife and Bootstrap.
Siegmund: Sequential Analysis: Tests and Confidence Intervals.
Simonoff: Smoothing Methods in Statistics.
Singpurwalla and Wilson: Statistical Methods in Software Engineering:
    Reliability and Risk.
Small: The Statistical Theory of Shape.
Sprott: Statistical Inference in Science.
Stein: Interpolation of Spatial Data: Some Theory for Kriging.
Taniguchi/Kakizawa: Asymptotic Theory of Statistical Inference for Time Series.
Tanner: Tools for Statistical Inference: Methods for the Exploration of Posterior
    Distributions and Likelihood Functions, 3rd edition.
Tong: The Multivariate Normal Distribution.
van der Vaart/Wellner: Weak Convergence and Empirical Processes: With
    Applications to Statistics.
Verbeke/Molenberghs: Linear Mixed Models for Longitudinal Data.
Weerahandi: Exact Statistical Methods for Data Analysis.
West/Harrison: Bayesian Forecasting and Dynamic Models, 2nd edition.